STABILITY AND POLITICIZATION IN CLIMATE GOVERNANCE

Tackling climate change requires long-term commitment to action, yet an array of influential parties with vested interests stand opposed to this. How best to engage and balance these positions for positive change is of increasing concern for advocates and policymakers. Exploring a discord within climate change policy and politics, this insightful volume critically examines the competing assumptions and arguments underpinning political "stability" versus "re/politicization" as a means of securing effective, long-term climate action. A range of cases exemplify the different political systems and power structures that underpin this antagonism, spanning geographical approaches, examples of nongovernmental action, and key industries in the global economy. Authored by an international team of scholars, this book will be of interest to researchers of local, national, and international legislation, specialists on climate governance policy, and other scholars involved in climate action. This title is also available as Open Access on Cambridge Core.

PAUL TOBIN is Associate Professor in Politics at the University of Manchester. He specializes in the politics and public policy of climate change, often within Europe. Paul has coedited two previous edited volumes, and two special issues, on environmental politics.

MATTHEW PATERSON is Professor of International Politics and Director of the Sustainable Consumption Institute at the University of Manchester. His research focuses on the political economy, global governance, and cultural politics of climate change. His latest book is *In Search of Climate Politics* (Cambridge University Press, 2021).

STACY D. VANDEVEER is Professor of Global Governance and Human Security in the John W. McCormack School of Policy and Global Studies at the University of Massachusetts Boston. He was 2023–24 Zennström Visiting Professor of Climate Leadership at Uppsala University and is a coeditor of *The Oxford Handbook of Comparative Environmental Politics* (Oxford University Press, 2023).

STABILITY AND POLITICIZATION IN CLIMATE GOVERNANCE

Edited by

PAUL TOBIN
University of Manchester

MATTHEW PATERSON
University of Manchester

STACY D. VANDEVEER
University of Massachusetts Boston

CAMBRIDGE
UNIVERSITY PRESS

CAMBRIDGE
UNIVERSITY PRESS

Shaftesbury Road, Cambridge CB2 8EA, United Kingdom

One Liberty Plaza, 20th Floor, New York, NY 10006, USA

477 Williamstown Road, Port Melbourne, VIC 3207, Australia

314–321, 3rd Floor, Plot 3, Splendor Forum, Jasola District Centre, New Delhi – 110025, India

103 Penang Road, #05–06/07, Visioncrest Commercial, Singapore 238467

Cambridge University Press is part of Cambridge University Press & Assessment, a department of the University of Cambridge.

We share the University's mission to contribute to society through the pursuit of education, learning and research at the highest international levels of excellence.

www.cambridge.org
Information on this title: www.cambridge.org/9781009352451

DOI: 10.1017/9781009352444

When citing this work, please include a reference to the DOI 10.1017/9781009352444

First published 2025

Cover image: Jeanne Rundell

A catalogue record for this publication is available from the British Library

Library of Congress Cataloging-in-Publication Data
Names: Tobin, Paul (Lecturer in politics), editor. | Paterson, Matthew, 1967– editor. | VanDeveer, Stacy D., editor.
Title: Stability and politicization in climate governance / edited by Paul Tobin, Matthew Paterson, Stacy D. VanDeveer.
Description: Cambridge ; New York, NY : Cambridge University Press, 2025. | Includes bibliographical references.
Identifiers: LCCN 2024058126 | ISBN 9781009352451 (hardback) | ISBN 9781009352444 (ebook)
Subjects: LCSH: Climatic changes – Government policy. | Climatic changes – Political aspects.
Classification: LCC QC903 .S72 2025 | DDC 363.7/0561–dc23/eng/20250314
LC record available at https://lccn.loc.gov/2024058126

ISBN 978-1-009-35245-1 Hardback

Contents

Figures

Tables

Contributors

Zarina Ahmad is a PhD scholar at the University of Manchester, UK, researching environmental sustainability and climate justice with underrepresented groups in society.

Nafhesa Ali is a lecturer in sociology at Northumbria University, UK, specializing in migration and the everyday lives of racialized and minority communities.

Jen Iris Allan is a senior lecturer at Cardiff University, UK. She specializes in global climate negotiations and activism. She has also published on biodiversity and hazardous chemicals and waste governance.

Steven Bernstein is Distinguished Professor of Global Environmental and Sustainability Governance at the University of Toronto, Canada. His research spans the areas of global governance and institutions, change and transformation in global politics, global environmental politics, international political economy, and policy studies.

Marc Debus is Professor of Comparative Government at the School of Social Sciences at the University of Mannheim, Germany. His research interests include political participation, party politics, coalition governance, legislative behavior, and decision-making in multilevel systems.

Fay M. Farstad is an associate professor in the Department of Comparative Politics at the University of Bergen, Norway, having been a senior researcher at the Center for International Climate Research (CICERO). Her research focuses on the party politics of climate change and EU climate policy.

Virginia Haufler is a professor in the Department of Government and Politics, University of Maryland, USA, and associate of the Center for International

Development and Conflict Management. She is a 2022–25 fellow of the Pierre Elliott Trudeau Foundation. Her research examines the relationship between private actors and public authority in global governance.

Erlend A. T. Hermansen is a senior researcher at the Center for International Climate Research (CICERO), Norway. He has a background in sociology and science and technology studies (STS) and his research revolves around how (scientific) knowledge is produced and used by different actors for different purposes across different contexts, including Norway.

Kathryn Hochstetler is Professor of International Development at the London School of Economics and Political Science, UK. She has published extensively on the relationship between environment and development. Her most recent book is *Political Economies of Energy Transition: Wind and Solar Power in Brazil and South Africa* (Cambridge University Press, 2021).

Matthew Hoffmann is Professor of Political Science and Codirector of the Environmental Governance Lab at the University of Toronto, Canada. His research and teaching interests include global climate politics, just transition, global environmental governance, and complex systems.

Bård Lahn is a postdoctoral fellow at the University of Oslo's Centre for Technology, Innovation and Culture, Norway, and an affiliated researcher at the Centre for International Climate Research (CICERO). With a background in STS, his work focuses on the politics of climate change, fossil fuels, and scientific expertise.

Joanna I. Lewis is Provost's Distinguished Associate Professor of Energy and Environment and Director of the Science, Technology and International Affairs Program at Georgetown University's Edmund A. Walsh School of Foreign Service, USA. She has more than two decades of experience working on international climate and clean energy policy with a focus on China.

Matthew Lockwood is a senior lecturer in energy policy at the Science Policy Research Unit, University of Sussex, UK. His main interests are in the politics of energy and climate policy, in the UK and comparatively.

Sherilyn MacGregor is Professor of Environmental Politics at the University of Manchester, UK, specializing in the politics of (un)sustainability, (in)equality, and environmental and climate justice.

Michael Méndez is an assistant professor of environmental planning and policy at the University of California, Irvine, USA, an Andrew Carnegie fellow, and a visiting scientist at the National Center for Atmospheric Research (NCAR). He previously served in California as a senior consultant, lobbyist, and gubernatorial appointee during the passage of the state's internationally acclaimed climate change legislation.

Matto Mildenberger is Assistant Professor at the University of California Santa Barbara (UCSB), USA, where he explores the comparative politics of climate change. He is the author of the book *Carbon Captured: How Business and Labor Control Climate Politics* (MIT Press, 2020). At UCSB, Matto directs the 2035 Initiative and coleads the Energy and Environment Transitions (ENVENT) Lab.

Jonas Nahm is an associate professor at the Johns Hopkins School of Advanced International Studies (SAIS) in Washington, DC, USA. His research interests lie in comparative political economy, at the intersection of climate policy, environmental politics, and economic and industrial policy.

Matthew Paterson is Professor of International Politics and Director of the Sustainable Consumption Institute at the University of Manchester, UK. His work focuses on the political economy, global governance, and cultural politics of climate change.

Wei Shen is a political economist who worked for development finance agencies in China for more than ten years. He is based at the Institute for Development Studies, which is affiliated with the University of Sussex, UK.

Martin Sokol is an economic geographer and associate professor at Trinity College Dublin, Ireland. His research focuses on geographies of finance, financialization, central banks, and climate justice. He is a founding member of the Global Network on Financial Geography (FinGeo) and Director of the Finance, Economy, Society Research Group.

Jennie C. Stephens is a feminist climate justice scholar-activist. She is Professor of Climate Justice at the National University of Ireland Maynooth, was a 2023–24 climate justice fellow at the Harvard Radcliffe Institute, and was previously a professor at Northeastern University, University of Vermont, USA, and Clark University, USA.

Yixian Sun is an associate professor in international development at the University of Bath, UK. He studies transnational governance, environmental politics, and sustainable development. His work explains China's changing role in global environmental governance, including sustainability transitions within China as well as sustainability impacts of China's overseas engagement.

Paul Tobin is a senior lecturer/associate professor in politics at the University of Manchester, UK. He specializes in the politics and public policy of climate change. In particular, Paul has researched the pathways to climate leadership as well as, in contrast, the processes that enable the dismantling of environmental legislation.

Diarmuid Torney is an associate professor in the School of Law and Government at Dublin City University (DCU), Ireland. His research focuses on climate change politics, policy, and governance. He is Director of the DCU Centre for Climate and Society and Chair of DCU's MSc in Climate Change: Policy, Media and Society.

Jale Tosun is Professor of Political Science at Heidelberg University, Germany, and Adjunct Professor at the University of Oslo, Norway. Her research concentrates on climate change governance, the European Union, policy studies, and public administration. She is the editor-in-chief of the Nature Portfolio journal *Climate Action*.

Stacy D. VanDeveer is Professor of Global Governance and Human Security at the John C. McCormack School of Policy and Global Studies at the University of Massachusetts Boston. He was 2023–24 Zennström Visiting Professor of Climate Change Leadership in the Department of Earth Sciences at Uppsala University, Sweden.

Acknowledgments

In February and March 2020, Mat and Paul were delighted to host Stacy as Hallsworth Visiting Professor at the University of Manchester. Mat and Paul held aspirations for holding a range of engaging workshops and guest lectures with Stacy regarding the latest in environmental politics. Sadly, a combination of the worst floods in years, strikes over pensions, and an impending accelerating pandemic rendered these events impossible. Nevertheless, we – Mat, Stacy, and Paul – did have a lovely walk in the Peak District and vowed to return to our initial discussions when context allowed.

Undeterred, we rearranged the research workshop as an online event, which was held across two days (a week apart) in October 2020. The vibrant and engaging discussions on attendees' ongoing research showed that the field of environmental politics was increasing straddling an antagonism about how to secure long-term climate action. We depicted this challenge by publishing a short Forum article in the journal *Global Environmental Politics* entitled "Climate Governance Antagonisms: Policy Stability and Repoliticization." The discussions at the online workshop were pivotal to the development of this research agenda, and we are most grateful to the attendees of the workshop, especially Elizabeth Bomberg, Charlotte Burns, Neil Carter, Brian Doherty, Robert Falkner, Andrew Jordan, Brendan Moore, Kate O'Neill, Rebecca Willis, Rüdiger Wurzel, and Anthony Zito, who were unable to participate in this subsequent edited volume but whose original insights are much appreciated.

Following positive responses to the initial Forum article, we proposed exploring this research agenda through an edited volume – the one you now read. To Matt Lloyd and Maya Zakrzewska-Pim at Cambridge University Press, thank you for your constant support for our project, for your patience as we navigated various deadlines and paperwork, and for your dedication to ensuring the resultant volume could be as strong as possible.

The support of the Economic and Social Research Council is gratefully acknowledged, having funded Paul Tobin via grant ES/S014500/1 during the editing of this book. Moreover, we are most appreciative of the UK Research and Innovation funding for Long-Form Outputs, which ensured that this volume could be published Open Access on Cambridge Core.

To the twenty-three outstanding scholars who joined us in this book project – thank you. We are grateful for your excellent work, engaging discussions, comments, and constructive critiques, and for your patience and good cheer.

We are extremely grateful to Mesilla, New Mexico-based artist Jeanne Rundell for allowing us to use her painting on the cover.

Paul, Mat, and Stacy are especially appreciative of the excellent research assistant work undertaken by Brian P. Harding, a doctoral candidate at the University of Massachusetts Boston, whose diligence in formatting the final manuscript was of the highest possible standard. Of course, any errors lie at the hands of the three editors.

Paul would like to thank Mat and Stacy for their mentorship and friendship, during this book project and beyond. Their combination of scholarship and kindness is a template that Paul will strive to emulate throughout his career. Further, the Department of Politics at the University of Manchester has been the ideal crucible in which to research environmental politics. Paul is grateful to his friends and family for their support throughout the editing of this book. Finally, as ever, from Paul to Ciara – thank you.

Mat would like to thank Paul and Stacy for being such good collaborators and friends. Paul did the lion's share of cat herding to keep us moving forward – thanks for that Paul! Mat has also benefited from regular conversations with a highly interdisciplinary group of colleagues at Manchester who work on various aspects of peat, led by the archaeologist Melanie Giles.

Stacy would like to thank Manchester University and its Department of Politics, and for support from the Hallsworth Visiting Professorship. He is also grateful for support from Uppsala University, its Department of Earth Sciences and its Climate Change Leadership (CCL) program – and to Niklas Zennström and the Zennström Visiting Professorship at CCL which he held for the 2023–24 academic year. He is also grateful to the University of Massachusetts Boston and his colleagues in the department of Conflict Resolution, Human Security and Global Governance for their tolerance of his inclinations to wander far afield. Lastly, many thanks to Paul and Mat for intellectual engagement and creativity and the good meals, long walks, park meet-ups, and occasional pub visits that make coauthoring and collaborating so much fun.

1

Introduction

Stability and Politicization in Climate Governance

MATTHEW PATERSON, PAUL TOBIN, AND STACY D. VANDEVEER

The early 2020s have been defined by endemic political instability, raising profound questions for the pursuit of climate policy. In many countries, seemingly stable climate policy regimes have been challenged by the rise of right-wing populism and various "backlash" movements, as in Sweden and the UK (Paterson et al. 2023; Patterson 2023). Meanwhile, in Hungary, Poland, and multiple states and provinces in the United States and Canada, climate obstruction remains a considerable threat to climate action. The geopolitical crisis provoked by the Russian invasion of Ukraine in 2022, coming quickly on the back of the COVID-19 pandemic, has generated novel pressures on climate policy due to myriad supply chain shocks, especially regarding natural gas, and the return of high inflation in many countries. This crisis has both empowered fossil fuel industries to seek support for increases in other sources of supply and created incentives to shift more rapidly toward renewables and electrification. Widening US–China conflict and competition continues to generate obstacles to global climate cooperation and incentives for trade as well as investment conflicts over green energy transitions. Chronic economic stagnation and widening economic inequality across many western countries, intensified by renewed inflation and economic (especially food and energy) insecurity, have fueled populist surges and resentments, framing climate action as unaffordable and elitist.

Although the last decade has seen many increasingly ambitious global, national, and subnational greenhouse gas (GHG) emissions reduction pledges – created around annual United Nations climate conferences, through European Union policymaking, and by myriad national governments and public and private transnational and subnational actors – a host of analyses continue to demonstrate that far too little in the way of emissions reductions has been achieved, and aggregate global emissions continue to rise (Anderson et al. 2020; Pörtner et al. 2022; Stoddard et al. 2021). This manifest failure of global and national political institutions to substantially and rapidly reduce emissions – most especially from high emitting countries, polities, and economic sectors – has added increasing fear, urgency, and rage to contemporary climate advocates, social movements, and critics of capitalism (Buzogány and Scherhaufer 2023; Malm 2021; de Moor 2023). Any previous ideas or expectations we may once have had that serious climate change policymaking by governments and states would result from greater scientific certainty about the causes of global climate change as well as increased public awareness and the visibility of climate change impacts can now be set aside. While much of this volume focuses on *mitigation* of climate change, the need for *adaptation* to its impacts grows more urgent – and expensive – by the day. Mitigation and adaptation are often championed by overlapping but distinct communities within global negotiations, and indeed, many scholars focus their research attentions on one or the other, but these two priorities are intricately and increasingly interconnected. As the need for greater adaptation efforts increases, so the requirement grows for already pledged funding from Global North countries to be shared as promised, yet current efforts remain woefully inadequate. As a result, inadequate and unfulfilled promises of support have escalated the urgency and anger surrounding the climate policy process.

These recent, multidirectional destabilizing dynamics have intensified an ongoing tension in climate policy and politics that we examine in this volume – namely, the antagonism between the pursuit of policy *stability* and the necessity of *(re) politicizing* climate change and its associated actors and institutions (Paterson et al. 2022).[1] What we call "stability" has long been a core goal among both policymakers and many academics studying climate policy. This pursuit of stability aims to create a policy regime that can secure long-term decarbonization as a response to climate change, in particular by insulating many core elements of the policy regime from the political competition that may undermine it. Just as Ulysses bound himself to his ship's mast, the goal of climate policy, as Levin et al. (2012: 123) put it, is to "constrain our future selves." But at the same time, many other commentators, analysts, and social movement activists (and the occasional politician) recognize that to achieve decarbonization it is essential to challenge the power of incumbent fossil fuel industries and their allies in political life. That is, we need more contestation, not less, and the pursuit of a stable policy environment may simply mean that the fossil fuel industry's power is never adequately undermined and is perhaps even strengthened.

This volume critically engages competing assumptions, strategies, and arguments underpinning stability "versus" conflict as a means of securing effective climate action. Both of these schools of thought enjoy many advocates within academia, while the stability argument is prevalent in the policy world, and the

[1] Parts of this introductory chapter draw heavily on Paterson et al. (2022).

politicization argument is more common among climate movement activists. However, as many of the contributions to this volume show, in practice, what exists is a more complex interplay between the two as climate policy develops in various contexts across the world. Moreover, in practice, each stance, as sketched in the second paragraph of this chapter, has in practice several important variations. Nevertheless, the dynamic between these two apparently contradictory imperatives in responding adequately to climate change shapes both empirically what we see happening in states, local governments, businesses, and movement campaigns and normatively how we think we should be continuing to imagine and advocate for more effective responses to the climate crisis.

This chapter is divided into three sections. First, we describe in more detail what we mean by stability. We explain the assumptions underpinning arguments for stability and then introduce four forms of stability that contributors to this volume employ. In the second section, we develop a threefold conceptualization of politics and use this to develop our account of re/politicization, similarly introducing four forms of re/politicization identified across the chapters in the book. Finally, we provide a thematic narrative of the arc of the chapters that follow.

1.1 Stability

Many policymakers proclaim a desire to create long-term climate policy stability. For instance, former European Union Commission officials Delbeke and Vis (2015: 2) argue that in designing climate policy, we need "to make sure that the policy context … remains as stable as possible." This perspective is embedded in many economists' discourses, most notably in arguments for the necessity of stable expectations about rising carbon prices and depoliticized mechanisms for ensuring that the control of such prices is outside the vagaries of politicians' intervention (Blyth et al. 2007; Bryant 2016; Jevnaker and Wettestad 2017), and thus means that for investors and other businesses, climate policy appears as a "credible commitment" (see Mildenberger and Lockwood, Chapter 13, this volume). Central bank independence is often taken as the paradigmatic case of this sort of policy stability (Stephens and Sokol, Chapter 7, this volume). At its strongest, the urge for policy stability can hold anti-political and anti-democratic impulses – as in James Lovelock's claim that "it may be time to put democracy on hold for a while" (in Willis 2020: 1).

But there is also an academic literature focused on "policy stability" that is only loosely connected to this immediate concern of policymakers. This literature proposes that institutions should be designed to ensure that governments embed climate policy across decision-making processes, seeking to "lock in" both those policies and institutions and to generate lock-in processes favoring low- or zero-carbon transformations. Such institutions seek to achieve this by, for example, establishing legal requirements for governments to achieve specific emissions reductions and new parastatal institutions that seek to monitor government policy. A paradigmatic example of "depoliticized" climate policy institutionalization in pursuit of climate policy stability is the UK's prescription in legislation of carbon budgets, the integration of climate policy at the cabinet level, and an expert-dominated and influential Climate Change Committee (Kuzemko 2016; Lockwood 2013; Willis 2020, see also Mildenberger and Lockwood, Chapter 13, and Torney, Chapter 12, this volume).

We see two principal sets of literature where this argument for policy stability is expressed. The first, developed among public policy analysts, focuses on detailed analysis of policy and institutional developments regarding climate change. This literature seeks to identify means by which policy design can automatically generate long-term effects that evade future backtracking, or at least create policy feedback mechanisms that hinder such backtracking progressively (Jordan and Moore 2020; Roberts et al. 2018; Rosenbloom et al. 2019). This research often stems from contexts of instability, such as following major crises, or where governance can be turbulent. Steinberg (2012), citing March and Olsen (2008), notes that a definitional value of a policy institution is its relative permanence.

A second set of literatures frames climate change as a "super-wicked problem," with conceptual foundations, such as path dependence, lock-in, and tipping points, originating in complex systems theory (Levin et al. 2012; Rietig and Laing 2017). Here, the concern for stability connects to the temporal dynamics of climate as a policy problem, in which political dynamics discount the future (Levin et al. 2012). The challenge is thus to "constrain our future selves" so that future politicians' instincts to discount the future cannot be expressed, thus generating a set of path dependencies toward low-carbon transition (see also Bernstein and Hoffmann, Chapter 16, this volume). Calls for climate change to be addressed through more "self-binding" legislation often build on other areas, such as monetary policy (Knaggård and Pihl 2015), while others seek to understand how to establish such forms of low-carbon path dependencies (Bernstein and Hoffmann 2018; Millar et al. 2020; Rosenbloom et al. 2019).[2]

We explored this tension in an earlier article (Paterson et al. 2022). As we developed the concept in this book, in dialogue with our various contributors, it became increasingly clear, however, that the meaning of stability in climate politics is,

[2] There is a literature that takes the logic of complex systems in a different direction, toward experimentation in climate governance (Bulkeley and Castán Broto 2013; Hoffmann 2011), where the logic in systems terms is that since no one knows a priori what combinations of emergent effects might generate tipping points toward low-carbon trajectories, climate policymakers should actively seek experimental initiatives.

Table 1.1 *The four forms of stability identified in this volume*

Form of stability	Appearance in the volume
Stability as the status quo	*Stability that perpetuates the structures and patterns that are already present.* This form of stability underpins more chapters than any other usage. For example, Tobin et al. (Chapter 4, this volume) explore how the status quo for Muslim climate actors in the UK is one of entrenched Islamophobia.
Stability as engineering lock-in	*Stability that depoliticizes via engineering solutions, without regard to politicization, and that builds on what came before.* This form of stability – manifesting through climate policy analysis that explicitly argues for such approaches and the assumptions of many policymakers who seek it – is an important background for the volume as a whole.
Stability as policy lock-in	*Stability that depoliticizes, as in the above, but via explicitly political means, and so with greater awareness of such dynamics as coalition building, policy feedbacks, and so on.* For example, Mildenberger and Lockwood (Chapter 13, this volume) analyze the pursuit of this form of stability in four countries.
Stability as long-term emissions reduction pathways	*Stability in the unwavering reduction of emissions, rather than a strategy for doing so.* While this form is an outcome, not a school of thought, and rarely appears explicitly in the book, it represents an important assumption as to the aspirations of those who pursue the other three forms of stability.

well, a little unstable. Across the volume, we encounter four principal forms of stability in climate politics, as outlined in Table 1.1.

As we develop these forms of stability, it becomes increasingly clear that a binary opposition between "stability" and "politicization" is overly simplistic. While stability as in *locked-in policy design* or *as engineering lock-in* can be seen as being depoliticized, and are often intended as such by their proponents, the other variants are not necessarily depoliticized; and even with those two, the picture becomes more complex. We will also see that there are complicated dynamics between these different senses of stability. For example, *stability as policy lock-in* is often promoted as a means of achieving engineering lock-in, or more simply as a means of generating smooth long-term emissions pathways. Yet the policy designs that result in reality may just as readily represent *capture* by fossil fuel interests, and thus end up as *stability as the status quo*, or at least in what Bernstein and Hoffmann (Chapter 16, this volume) call "improvement": a transition away from the original high-carbon situation but which locks in still inadequate responses. In other contexts, pursuing *stability as long-term emissions reduction pathways* (e.g., via a carbon budget regime) in the end entails what Paterson (Chapter 6, this volume) calls "scope expansion," introducing new sources of emissions or sinks into the policy and political debate which then risk destabilizing the policy regime because of the way those new sources generate political conflict.

1.2 Re/politicization

Despite the potential benefits of stability, its antagonists propose that there are significant dangers in prioritizing the pursuit of climate policy stability. The normative priority of stability entails climate policy that avoids conflicts. A common critique of prioritizing policy stability is that it is either naïve about or designed to avoid conflict with entrenched interests who gain from the perpetuation of the status quo. Existing policy regimes have been highly stable in the way they have failed to decarbonize societies in any sustained way precisely *because* existing stable policy regimes entrench fossil fuel interests and power. The argument by academics in favor of "policy stability" to promote decarbonization can therefore readily be co-opted by the policymakers whose pursuit of policy stability – or quite simply, other ends – is deeper than their commitment to decarbonization. For these analysts, "the proof is in the pudding" as three decades of climate policymaking have failed to bend the global GHG emissions curve and barely two dozen countries have seen sustained emissions declines, with most of these quite modest indeed (Lamb et al. 2021; Stoddard et al. 2021).

Alongside this normative/strategic critique, it is an empirical mistake to assume that climate policy can be addressed "outside politics." Politics – when understood as conflicts of interest and power, and the ongoing necessity of collective decision-making – is simply intrinsic to social life. Thus, the pursuit of stability produces a second sort of danger: The avoidance of political conflicts can itself bite back. This threat is demonstrated most obviously by the persistence of climate denial (Brulle 2018; Jacques 2023; Supran and Oreskes 2017) in numerous – often fossil fuel-funded – contexts, including in details of rulemaking, standard-setting, and regulatory design, which offer innumerable opportunities for powerful interests to resist and limit the very changes policies are designed to make (Stokes 2020). Another manifestation of this "backlash" (Patterson 2023) is that of popular resistance to climate policy, such as opposition to the development of carbon pricing in many neoliberal states (MacNeil 2016), or what occurs once carbon pricing is initiated, as in the case of the French *gilets jaunes* (Gaborit and Grémion 2019), or once policies have previously been "successfully" introduced, as in the more recent populist turns in UK and Swedish climate politics (Paterson, Wilshire, and Tobin 2024; Vihma, Reischl, and Andersen 2021).

Stability in the sense of locked-in policy design on its own cannot be a guarantee of the rapid and profound transformations we need. The literature on politicization revives a longer history of analysis of corporate actors in climate politics (Jordan and Moore 2020; Levy and Newell 2005; Meckling 2011; Newell and Paterson 1998), as well as work on climate denial emphasizing the role of corporations (Brulle 2018; Supran and Oreskes 2017). Periodically, climate scientists and NGO activists discuss how corporations have undermined climate policy (Leggett 1999; Mann 2021). These arguments are connected to a broader set of what might be called "left populist" arguments among climate justice writers and activists, exemplified by Klein (2014) and McKibben (2012). McKibben frames the fossil fuel industry as the "enemy," allying with arguments that note how there are "[j]ust 100 companies responsible for 71% of global emissions" (Riley 2017). Allan (2021) shows that a growing coalition of "radical challengers" espousing anti-capitalist and anti-establishment views are increasingly strongly criticizing the United Nations' climate governance system, the "traditional" state-based and environmental nongovernmental organizations (NGOs) structuring it, and the pro-market policy instruments they promulgate. These left populist campaigns have targeted variously oil pipelines, fracking, and divestment to contest corporate power but also opened up political debate by promoting strategies centered on the idea of a Green New Deal.

Eco-Marxist analyses often, although not always, underpin these repoliticization efforts. Malm's meticulous (2015) demonstration of how industrialization was driven by capitalist logics of control over labor and inter-capitalist competition, rather than explanations around coal being more efficient, or similar economistic rationales, has become a classic such analysis. In doing so, he and others have connected populist anti-corporate arguments to systemic understandings about the character of capitalist development. Further, Mouffe (2005) pitches the argument in general, but it resonates particularly for climate politics given the depth of social transformations required and the deep incumbent interests threatened by such changes (Paterson 2020). Recent related arguments have been advanced as critiques of most of the literature on global governance, thus appreciating the value of politicizing global governance institutions (Acharya 2018; Zurn 2018). Related work can also politicize governance practices, priorities, and discourses – and their ideological bases – at the same time, as Maechler and Boisvert (2024) do in their critique of the hegemonic "valuation-centrism" on contemporary nature and biodiversity governance initiatives.

As economic, environmental, and climate change inequalities worsen from the global to local scale, attention to justice and injustice in the discourses and governance practices of contemporary climate change politics and research has also increased. Research focused on various scales of climate justice politics and research has also grown substantially (Flavell 2024; Newell et al. 2021; Shi et al. 2016; Sowers, VanDeveer, and Weinthal 2023; Tobin et al. 2024; Zimm et al.

2024). Such work aligns with the view that existing injustices necessitate politicization in research, scholarship, policy, and governance practice. In other words, if racism, gender biases, neoliberalism, and/or colonialism are "institutionalized" in past and present governance practices and discourses – as such literatures tend to either assume (ideologically) or demonstrate (empirically) – then demands for and attempts at change are either inherently or likely to be framed as political (see Christophers 2024; Paterson 2024).

Underpinning these different uses of the term "politicization" (and indeed, "stability") is a more basic question of what we mean by "politics." Politics is famously an "essentially contested concept" (Gallie 1956), meaning both that it is impossible to arrive at a definition that all can agree on, but also that the choices of usage are themselves political in their implications. In this context, we think it useful to think of politics as three interacting dimensions, as illustrated in Figure 1.1 (for details, see Paterson 2021). First, politics is a particular type of site or arena: one where authoritative collective decisions are made, entailing particular types of processes – more or less deliberative or democratic – to generate these decisions. This understanding is often but not necessarily reduced to the state as the principal site of politics. Second, politics is about the question of power in social life – how it is generated, distributed, exercised, and legitimized. Third, politics is intrinsically conflictual: Societies of any complexity entail deeply structured inequalities of power and wealth, as well as competing visions and values, in ways that mean that arriving at collective authoritative decisions is rarely consensual, or at least conflicts will be recurrent.

Given this threefold understanding of politics, depoliticization – as sought by those pursuing stability as policy lock-in or as engineering lock-in, most notably – can be understood as a process whereby actors seek to take a phenomenon (climate action) out of formal political arenas (i.e. ones with some semblance of open deliberative decision-making, although not necessarily democratic) into ones without that quality (technocratic, etc.). They do so normally as a strategy of power, where

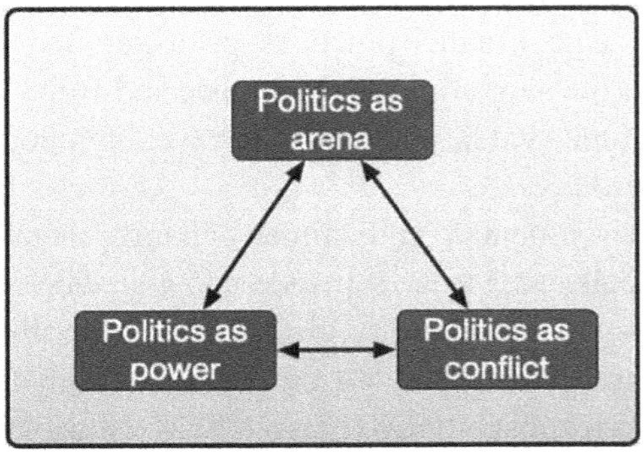

Figure 1.1 Three dimensions of politics (Paterson 2021).

powerful actors seek to insulate themselves from accountability – as we see in many of our chapters (i.e. VanDeveer, Chapter 5; Haufler, Chapter 8; Stephens and Sokol, Chapter 7; Paterson, Chapter 6). But occasionally depoliticization has other dimensions, for example where marginalized groups may seek to depoliticize their climate interventions to avoid the potential of making themselves vulnerable via their activism (see Tobin et al., Chapter 4, this volume).

Hence, repoliticization is largely the reverse process to that of depoliticization – that is, seeking to bring processes back into explicitly political arenas in order to open up and contextualize existing power relations, thus making conflicts of power and interests explicit. Depoliticization is "orthogonal" to stability – stability is often but not always depoliticizing, and there is a subtle set of relations shaping the interactions between the two concepts.

Again, as with stability, it became clear as we explored politicization with our contributors that there are multiple forms of the strategy, including myriad implications of the prefixes de- and re- before any politicization. Existing

Table 1.2 *The four forms of politicization identified in this volume*

Form of politicization	Appearance within the volume
Politicization as broader sociopolitical change	*Politicization that disrupts the* status quo, *beyond direct impacts on climate change.* For example, Méndez (Chapter 3, this volume) examines coalitions of actors in California that pursue justice across myriad areas.
Politicization as partisan competition	*Politicization that elicits competition, be it polarizing politicization (pro- and anti-climate policy), or politicization of performance (regarding who has the better climate policy).* At the national level, we see these dynamics at play across four countries in the chapter by Mildenberger and Lockwood (Chapter 13, this volume); at the sectoral level it manifests in those by VanDeveer (Chapter 5), by Paterson (Chapter 6), and by Haufler (Chapter 8); while at the global level, Allan (Chapter 14), and Nahm (Chapter 15) both examine geopolitical competition.
Politicization as discourse	*Politicization that functions as a discursive tool, be it a process that brings its object into political debate, a strategy of actors to seek to do so, or as a means of shifting attention onto rhetoric rather than activity.* For example, Allan (Chapter 14) notes rhetorical efforts to repoliticize issues within global climate negotiations.
Politicization as scholarly praxis	*Politicization that arises through scholarly interventions.* For example, several chapters function in this manner, most explicitly Tobin et al.'s (Chapter 4) efforts to lessen the invisibilization of Muslim communities in policymaking and within academic literature.

scholarship has varied in its focus between politicization and *re*politicization. The main advantage of the prefix re- is that repoliticization emphasizes that it is an iterative and relational dynamic between re- and de-. The logic of this distinction is that there is no such thing as simply politicization, since this dynamic is always ongoing, and there is no fictional moment "before politics." In this volume, we are agnostic about the usage of politicization and repoliticization. In much of the literature, politicization and repoliticization are treated as being the same thing, and so we look to our contributing authors to decide which term is most apt for their chapter.

Table 1.2 introduces the four forms of politicization we have identified in this book.

1.3 The Volume's Structure: From Movement Activism to Global Politics

This book is divided into five parts. We start by focusing on *social movement activism*. Civil society and non-state action has long been a realm of climate action because of the establishment of fossil fuel interests within state infrastructure, necessitating climate movements to find their voices "outside" government. Movements are often key actors in seeking to politicize climate action, by contesting the power of incumbent actors that seek to block such action, and thus in our sense of pursuit of broader sociopolitical change, and relatively rarely in the sense of immediate partisan competition among political parties. Tosun and Debus (Chapter 2) thus focus on perhaps the most visible global forms of such movement action in recent years: the Fridays for Future school strikes movement. They show that this activism has opened up new conflicts in German climate policy and forced the German government to shift to make various aspects of its climate strategy more ambitious. Méndez (Chapter 3) shifts the focus to climate justice movements, focused on California, and the way that individuals and organizations are dynamically co-constituting what climate justice means through ongoing conflict and collaboration. In contrast perhaps, Tobin and coauthors (Chapter 4) show how in some contexts – such as their case of Muslim climate actors in the UK – movements may wish to see politicization in order to disrupt the status quo but are in a vulnerable position for securing these changes due to Islamophobia and hence look beyond protests to secure politicization, and prioritize wider sociopolitical justice that is interwoven with climate justice.

In Part II, we turn our focus to a frequent target of movement attempts to politicize climate change: key industries in the global economy. Hence, we explore the *political economy dimensions of climate change* central to many arguments for repoliticization. We thus focus variously on the dynamics of coal phaseouts across the world (VanDeveer, Chapter 5) and the contentious politics and iterated politicization they engender, the politics of land ownership and peat moor management in the UK (Paterson, Chapter 6), the recent involvement in climate politics by central banks (Stephens and Sokol, Chapter 7), and the usage of voluntary climate initiatives within insurance industry as a means of pursuing depoliticization (Haufler, Chapter 8). In most of these chapters, we see that stability instead ends up entrenching the status quo because of prevailing power relations. But we also see some of the contradictions of stability-oriented climate policy as it unfolds, as in Paterson's argument that climate policy brought new objects – peat moors – into view that carried with them preexisting political relations and conflicts. Yet these political economy dimensions, and the sorts of pressure movements can bring to bear, vary across countries.

Part III thus focuses on key dimensions of *comparative climate policy*. We start with Hochstetler's chapter (Chapter 9), where the underlying political economies of Brazil and South Africa shape the patterns of stability and politicization that unfold in each. In Brazil, electricity is not dominated by fossil fuel interests, while there is fierce contestation over the question of deforestation. Resultantly, there is a relatively depoliticized climate strategy (including being supported by right-wing populists like Bolsonaro) focused on wind energy development. By contrast, in South Africa, with electricity publicly owned and dominated by coal interests, stability has meant the *status quo*, and pushing for renewable energy has been highly contentious. Sun, Shen, and Lewis (Chapter 10) focus on China, showing how, despite very strong incumbent fossil fuel actors, the state's coordinating authority has managed to generate policy to accelerate the uptake of renewables, without significant explicit politicization – a noteworthy finding considering that much of the existing literature focuses on democracies' climate strategies. Farstad, Hermansen, and Lahn (Chapter 11) focus on the Norwegian case – a classic "consensual" decision-making system, showing that even there a series of specific dynamics in the Norwegian economy and party system have led to such a depoliticized arrangement coming under significant pressure as climate policy has unfolded. Torney (Chapter 12) turns his attention more explicitly to the stability argument in relation to its common referent – overarching climate laws usually understood as aiming to "lock in" decarbonization – and shows the picture is somewhat more complicated, with many such laws reflecting political compromises but more importantly designing in a recognition of future political conflicts. Mildenberger and Lockwood (Chapter 13) similarly analyze classically "stability-oriented" climate strategies but focus more on how, in practice, cases reflect explicitly political bargains or party competition within each case.

Part IV responds to the reality that the literature on stability and politicization has mostly focused on national policymaking processes, by examining *global politics.* The existing focus on states partly derives from the implicit argument within pieces favoring stability regarding the sovereign state being able to authoritatively impose consistent policy and strategy across time. But, of course, global political dynamics are both key to climate change and have no overall sovereign actor. Various chapters have already hinted at such international or transnational dynamics, notably VanDeveer (Chapter 5) on coal phaseout, Stephens and Sokol (Chapter 7) on central bank coordination, and Haufler (Chapter 8) on the global insurance industry. But in Part IV, we focus explicitly on what happens to the stability/politicization argument if we discuss it in global political contexts, with the attendant dynamics that underpin specifically global interactions. Allan (Chapter 14) focuses on the core of the international climate regime – the United Nations Framework Convention on Climate Change (UNFCCC) – and argues that we can think of this dynamic in terms of attempts to create long-lasting treaty designs, which have been periodically repoliticized for reasons of geopolitical strategy. Specifically, the Kyoto Protocol embedded a design for what were intended as durable institutions, but these institutions rapidly became repoliticized by the United States, which decided that the Kyoto design (despite its principal features having US origins) no longer served US geopolitical interests. Nahm (Chapter 15) switches attention to the question of global trade and industrial competition. He uses this sectoral case to show how the stability/politicization dynamic becomes complicated when we think between national and international spheres of political life. Nahm argues that states may promote renewable energy development as a means of building domestic coalitions of support for investment in manufacturing that in turn generate "policy feedback." However, this pursuit of domestic political stability risks triggering international repoliticization centered on the rules of the global trade regime, thus demonstrating the delicate interconnectedness of stability and politicization.

We begin the concluding Part V of the book with Bernstein and Hoffmann's (Chapter 16) reformulation of what they have elsewhere (Bernstein and Hoffmann 2019) called the "carbon trap." While the idea of the trap is couched in the theoretical language of complex systems that informs the stability approach in particular, Bernstein and Hoffmann extend this conceptualization here by considering politicization as integral to climate policy processes. They use this understanding to reformulate the model by showing how the processes of incumbent resistance, political compromise, and coalition building can result in what they call "improvement" – a situation where societies are shifted from high carbon but end up locking in technologies and investments that prevent full decarbonization. For example, fossil fuel interests may be able to capture the process to mean that hydrogen (from natural gas) and carbon capture and storage become regarded as "solutions." Finally, in the concluding chapter of the book, VanDeveer, Tobin, and Paterson (Chapter 17) seek to synthesize the arguments made within the volume, drawing particular focus onto the *agents* behind the dynamics of stability and politicization, and calling for future avenues of research that build on this book's findings.

References

Acharya, A. (2018). *Constructing Global Order: Agency and Change in World Politics*. Cambridge: Cambridge University Press.

Allan, J. I. (2021). *The New Climate Activism: NGO Authority and Participation in Climate Change Governance*. Toronto: University of Toronto Press.

Anderson, K., Broderick, J., and Stoddard, I. (2020). A factor of two: How mitigation plans of "climate progressive" nations fall short of Paris-compliant pathways. *Climate Policy* 20(10), 1290–1304.

Bernstein, S., and Hoffmann, M. (2018). The politics of decarbonization and the catalytic impact of subnational climate experiments. *Policy Sciences* 51(2), 189–211.

Bernstein, S., and Hoffmann, M. (2019). Climate politics, metaphors and the fractal carbon trap. *Nature Climate Change* 9(12), 919–925.

Blyth, W., Bradley, R., Bunn, D. et al. (2007). Investment risks under uncertain climate change policy. *Energy Policy* 35(11), 5766–5773.

Brulle, R. J. (2018). The climate lobby: A sectoral analysis of lobbying spending on climate change in the USA, 2000 to 2016. *Climatic Change* 149(3), 289–303.

Bryant, G. (2016). The politics of carbon market design: Rethinking the techno-politics and post-politics of climate change. *Antipode* 48(4), 877–898.

Bulkeley, H., and Castán Broto, V. (2013). Government by experiment? Global cities and the governing of climate change. *Transactions of the Institute of British Geographers* 38(3), 361–375. https://doi.org/10.1111/j.1475-5661.2012.00535.x

Buzogány, A., and Scherhaufer, P. (2023). The new climate movement: Organization, strategy, and consequences. In H. Jörgens, C. Knill, and Y. Steinebach, eds., *Routledge Handbook of Environmental Policy*. Abingdon: Routledge, pp. 358–380.

Christophers, B. (2024). *The Price Is Wrong: Why Capitalism Won't Save the Planet*. London: Verso Books.

Delbeke, J., and Vis, P., eds. (2015). *EU Climate Policy Explained*. Abingdon: Routledge.

Flavell, J. (2024). From gender-blind to gender bind: Foregrounding gender in the history of the UNFCCC. *Global Environmental Politics* 24(1), 31–51.

Gaborit, M., and Grémion, T. (2019). Jaunes et verts. *La Vie des idées*. https://laviedesidees.fr/Jaunes-et-verts.html.

Gallie, W. B. (1956). Essentially contested concepts. In M. Black, ed., *The Importance of Language*. Ithaca, NY: Cornell University Press, pp. 121–146.

Hoffmann, M. (2011). *Climate Governance at the Crossroads: Experimenting with a Global Response after Kyoto*. Oxford: Oxford University Press. https://doi.org/10.1093/acprof:oso/9780195390087.001.0001

Jacques, P. (2023). Anglo fears: Rejection of climate change and Anglo anxiety. In J. Sowers, S. D. VanDeveer, and E. Weinthal, eds., *Oxford Handbook of Comparative Environmental Politics*. Oxford: Oxford University Press, pp. 335–355.

Jevnaker, T., and Wettestad, J. (2017). Ratcheting up carbon trade: The politics of reforming EU emissions trading. *Global Environmental Politics* 17(2), 105–124.

Jordan, A., and Moore, B. (2020). *Durable by Design? Policy Feedback in a Changing Climate*. Cambridge: Cambridge University Press.

Klein, N. (2014). *This Changes Everything: Capitalism vs. The Climate*. Toronto: Knopf Canada.

Knaggård, Å., and Pihl, H. (2015). The green state and the design of self-binding. In K. Bäckstrand and A. Kronsell, eds., *Rethinking the Green State: Environmental Governance towards Climate and Sustainability Transitions*. Abingdon: Routledge, pp. 209–224.
Kuzemko, C. (2016). Energy depoliticisation in the UK: Destroying political capacity. *The British Journal of Politics and International Relations* 18(1), 107–124.
Lamb, W. F., Grubb, M., DiLuiso, F., and Minx, J. (2021). Countries with sustained greenhouse gas reductions. *Climate Policy* 22(1), 1–17.
Leggett, J. (1999). *The Carbon War: Dispatches from the End of the Oil Century*. London: Penguin.
Levin, K., Cashore, B., Bernstein, S., and Auld, G. (2012). Overcoming the tragedy of super wicked problems: Constraining our future selves to ameliorate global climate change. *Policy Sciences* 45(2), 123–152.
Levy, D., and Newell, P., eds. (2005). *The Business of Global Environmental Governance*. Cambridge, MA: MIT Press.
Lockwood, M. (2013). The political sustainability of climate policy: The case of the UK Climate Change Act. *Global Environmental Change* 23(5), 1339–1348.
MacNeil, R. (2016). Death and environmental taxes: Why market environmentalism fails in liberal market economies. *Global Environmental Politics* 16(1), 21–37. https://doi.org/10.1162/GLEP_a_00336
Maechler, S., and Boisvert, V. (2024). Valuing nature to save it? The centrality of valuation in the new spirit of conservation. *Global Environmental Politics* 24(1), 10–30.
Malm, A. (2015). *Fossil Capital: The Rise of Steam-Power and the Roots of Global Warming*. London: Verso Books.
Malm, A. (2021). *How to Blow Up a Pipeline*. London: Verso Press.
Mann, M. E. (2021). *The New Climate War: The Fight to Take Back Our Planet*. New York: Public Affairs.
March, J. G., and Olsen, J. P. (2008). Elaborating the "new institutionalism." In R. A. W. Rhodes, S. Binder, and B. Rockman, eds. *The Oxford Handbook of Political Institutions*. Oxford: Oxford University Press, pp. 3–22.
McKibben, B. (2012). Global warming's terrifying new math. *Rolling Stone*, July 19. www.rollingstone.com/politics/news/global-warmings-terrifying-new-math-20120719.
Meckling, J. (2011). *Carbon Coalitions: Business, Climate Politics, and the Rise of Emissions Trading*. Cambridge, MA: MIT Press.
Millar, H., Bourgeois, E., Bernstein, S., and Hoffmann, M. (2020). Self-reinforcing and self-undermining feedbacks in subnational climate policy implementation. *Environmental Politics* 30(5), 791–810.
de Moor, J. (2023). Time and place in climate activism: Three urgency-induced debates. In J. Sowers, S. D. VanDeveer, and E. Weinthal, eds., *Oxford Handbook of Comparative Environmental Politics*. Oxford: Oxford University Press, pp. 299–316.
Mouffe, C. (2005). *On the Political* (new ed.). London: Routledge.
Newell, P., and Paterson, M. (1998). A climate for business: Global warming, the state and capital. *Review of International Political Economy* 5(4), 679–703.
Newell, P., Srivastava, S., Naess, L. O., Torres Contreras, G. A., and Price, R. (2021). Toward transformative climate justice: An emerging research agenda. *Wiley Interdisciplinary Reviews: Climate Change* 12(6), e733.
Paterson, M. (2020). Climate change and international political economy: Between collapse and transformation. *Review of International Political Economy* 28(2), 394–405.
Paterson, M. (2021). *In Search of Climate Politics*. Cambridge: Cambridge University Press.
Paterson, M. (2024). An unjust transition. *The Ecologist*, February 12. https://theecologist.org/2024/feb/12/unjust-transition.

Paterson, M., Tobin, P., and VanDeveer, S. D. (2022). Climate governance antagonisms: Policy stability and repoliticization. *Global Environmental Politics* 22(2), 1–11.

Paterson, M., Wilshire, S., and Tobin, P. (2024) The rise of anti-net zero populism in the UK: Comparing rhetorical strategies for climate policy dismantling. *Journal of Comparative Policy Analysis: Research and Practice* 26(3–4), 332–350.

Patterson, J. J. (2023). Backlash to climate policy. *Global Environmental Politics* 23(1), 68–90.

Pörtner, H. O., Roberts, D. C., Poloczanska, E. S. et al. (2022). *Summary for Policymakers*. Contribution of Working Group II to the Sixth Assessment Report of the Intergovernmental Panel on Climate Change. Cambridge: Cambridge University Press. https://doi.org/10.1017/9781009325844.001.

Rietig, K., and Laing, T. (2017). Policy stability in climate governance: The case of the United Kingdom. *Environmental Policy and Governance* 27(6), 575–587.

Riley, T. (2017). Just 100 companies responsible for 71% of global emissions, study says. *The Guardian*, July 10. www.theguardian.com/sustainable-business/2017/jul/10/100-fossil-fuel-companies-investors-responsible-71-global-emissions-cdp-study-climate-change.

Roberts, C., Geels, F. W., Lockwood, M. et al. (2018). The politics of accelerating low-carbon transitions: Towards a new research agenda. *Energy Research & Social Science* 44, 304–311.

Rosenbloom, D., Meadowcroft, J., and Cashore, B. (2019). Stability and climate policy? Harnessing insights on path dependence, policy feedback, and transition pathways. *Energy Research & Social Science* 50(April), 168–178.

Shi, L., Chu, E., Anguelovski, I. et al. (2016). Roadmap towards justice in urban adaptation research. *Nature Climate Change* 6, 131–137.

Sowers, J., VanDeveer, S. D., and Weinthal, E., eds. (2023). *Oxford Handbook of Comparative Environmental Politics*. Oxford: Oxford University Press.

Steinberg, P. F. (2012). Welcome to the jungle: Policy theory and political instability. In S. D. VanDeveer and P. F. Steinberg, eds., *Comparative Environmental Politics: Theory, Practice, and Prospects*. Cambridge, MA: MIT Press, pp. 255–284.

Stoddard, I., Anderson, K., Capstick, S. et al. (2021). Three decades of climate mitigation: Why haven't we bent the global emissions curve? *Annual Review of Environment and Resources* 46, 653–689.

Stokes, L. C. (2020). *Short Circuiting Policy: Interest Groups and the Battle over Clean Energy and Climate Policy in the American States*. Oxford: Oxford University Press.

Supran, G., and Oreskes, N. (2017). Assessing ExxonMobil's climate change communications (1977–2014). *Environmental Research Letters* 12(8), 084019.

Tobin, P., Ali, N., MacGregor, S., and Ahmad, Z. (2024). "The religions are engaging: Tick, well done": The invisibilization and instrumentalization of Muslim climate intermediaries. *Policy Studies* 44(5), 627–645.

Vihma, A., Reischl, G., and Nonbo Andersen, A. (2021). A climate backlash: Comparing populist parties' climate policies in Denmark, Finland, and Sweden. *The Journal of Environment & Development* 30(3), 219–239.

Willis, R. (2020) *Too Hot to Handle? The Democratic Challenge of Climate Change*. Bristol: Policy Press.

Zimm, C., Mintz-Woo, K., Brutschin, E. et al. (2024). Justice considerations in climate research. *Nature Climate Change* 14, 22–30.

Zurn, M. (2018). *A Theory of Global Governance: Authority, Legitimacy and Global Governance*. Oxford: Oxford University Press.

Part I

Movement Politics

2

The Fridays for Future Movement and the Repoliticization of Climate Change Policy in Germany

JALE TOSUN AND MARC DEBUS

Germany is among the most studied countries in environmental and climate politics due to the distinct features of its political system and some pioneering policy decisions made in the past, such as the feed-in electricity tariffs to encourage the use of new renewable energy technologies. One of the most outstanding features of the German political system is the sustained electoral strength of its green political party, Alliance '90/The Greens (henceforth: the Greens) since the 1980s. The Greens served as a junior coalition partner in two federal governments between 1998 and 2005 and are members of the current federal government led by Chancellor Olaf Scholz. Since 2011, the Minister-President of the State of Baden-Württemberg has been the Green politician Winfried Kretschmann, and in many other states the Greens have participated in government as junior coalition partners since the mid 1980s.

As the Greens "own" the issue of environmental protection (Abou-Chadi 2016), they are the obvious electoral preference of Germans who are concerned about the environment. The other political parties are not against environmental protection, but they give less priority to it than the Greens (Carter 2013). Since all German governments, regardless of their ideological composition, have adopted public policies addressing environmental degradation, environmental issues have been of mostly moderate-level salience: Citizens were aware of policy activity and this was enough for most of them.

However, this changed in two ways in 2018. First, the salience (i.e. the awareness of the citizens) of environmental issues increased dramatically; second, the public's attention shifted away from environmental protection to tackling climate change, which until then had been perceived as a mere aspect of environmental protection and consequently had received little attention as a stand-alone issue, including from political parties (Farstad 2018).

What resulted in this shift in the German public's awareness of climate change? In August 2018, the then teenager Greta Thunberg boycotted school and instead protested in front of the Swedish parliament to call for more urgent and effective climate action. The protest rapidly attracted followers, giving birth to the Fridays for Future (FFF) movement, which borrowed its name from the corresponding social media hashtag (Venghaus, Henseleit, and Belka 2022). As in other countries, and despite the fact that Germany has an electorally successful Green party, FFF succeeded in mobilizing a large number of individuals, of whom most were newcomers to social movements, such as pupils (Moor et al. 2021). Given the mobilization of young people in particular, FFF credibly made a case for the generational aspect of anthropogenic greenhouse gas (GHG) emissions and the need for urgent climate action in order to protect the interests of current young and future generations (Tosun, Geese, and Lorenzoni 2023).

The mobilization of this group was unprecedented, even in Germany, and resulted in solidarity and the launch of spin-off movements. The involvement of these groups in climate activism has helped to draw even broader attention to the issue of climate change and resulted in an increased salience of climate change.

In this chapter, we hypothesize increased salience to have resulted in a politicization of the issue and affected climate politics in general and party competition more specifically. In the remainder, we assess this argument empirically and show that Germany has embarked on a journey to reduce carbon emissions from burning coal and is in the process of changing the status quo of its climate policy.

2.1 Is Climate Change Salient and Politicized?

Politicization, defined in different ways (see Paterson, Tobin, and VanDeveer, Chapter 1, this volume), has long been a research interest of political science and has been studied in relation to electoral campaigns, voting behavior, and parliamentary representation. An increasing number of empirical studies have applied this notion to European Union (EU) politics following a rise in the number of legal competences transferred to EU institutions (Grande and Hutter 2016; Wilde and Zürn 2012). Recently, politicization has also entered policy studies. In this context, Feindt, Schwindenhammer, and Tosun (2021), for instance, contend that political action can be conducive to policy change as long as it does not surpass a “critical” level.

In the literature, the conceptualization of politicization by Wilde (2011) has received considerable attention. Wilde argues that politicization is observable in rising salience, mobilization, and polarization. Consequently, salience is only one of three indicators and does not alone account for politicization. Mobilization, the second indicator, refers to an increase in the number of political actors engaged in an issue and the resources they spend on their engagement. The third indicator, polarization, denotes the emergence of (extremely) opposing demands in addressing the issue concerned (Wilde and Zürn 2012).

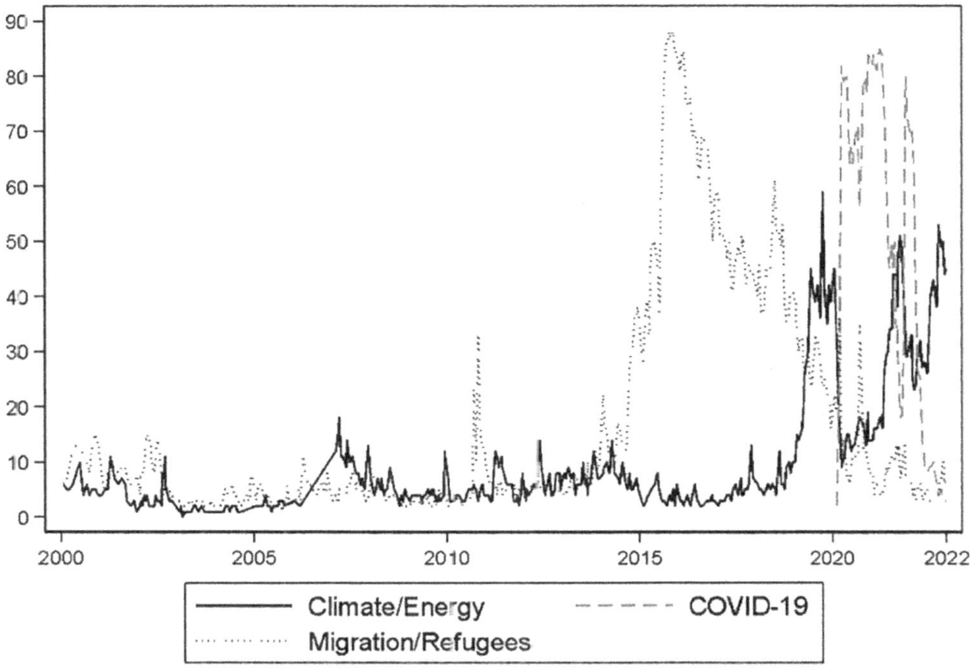

Figure 2.1 Percentage share of individuals indicating climate and energy policy as the most important problem in Germany, 2000–2022.
Note: Own elaboration based on data from the Research Group for Elections, www.forschungsgruppe.de (the full data are available from Forschungsgruppe Wahlen E.V. 2024).

As stated in the opening section of this chapter and argued consistently in the literature, the emergence of FFF has increased the salience of climate change. The data from the 2022 Politbarometer survey confirm this assessment. The Politbarometer has been conducted since 1977 on an almost monthly basis by the Research Group for Elections. The survey focuses on the opinions and attitudes of the German voting population on current political topics, parties, and politicians, and it screens the voting intention of the respondents. It also provides an apt database for learning about the salience of climate and energy issues and how they have developed over time. The survey item capturing an issue's salience is the problem which the survey participants consider to be of the highest importance in Germany. Figure 2.1 illustrates the salience of energy and climate issues, which were treated as one joint category, between 2000 and 2022.

The data show that there was some concern throughout the 2000s and 2010s over climate and energy issues, but it was volatile. A well-known event that drove up public concern among Germans regarding climate and energy was the accident that occurred on March 11, 2011, at the Fukushima Daiichi Nuclear Power Plant in

Japan. This event marked a turning point for nuclear power production in particular and the energy policy of the federal government in general (see, e.g., Jahn and Korolczuk 2012). Attention to climate and energy issues went down in the second half of the 2010s because of the massive influx of refugees to Germany and the EU in 2015 and 2016, which became the dominant concern among Germans. However, in 2019 climate concern skyrocketed, which coincides with the emergence of FFF. The outbreak of COVID-19 then became the main concern in 2020. In 2021 and 2022, both issues competed against each other for attention. The last data point captured by Figure 2.1 for the year 2022 shows that climate change was on the rise again, whereas COVID-19 was of minimal concern to Germans, which is plausible given the wide availability of vaccines and the emergence of milder virus variants.

The second indicator of politicization, mobilization, is given because of the emergence and institutionalization of FFF, though it should be noted that FFF was not the only social movement targeting climate change which emerged around the same time. A second emblematic group is Extinction Rebellion, which has equally mobilized large numbers of people, particularly young people, since it was founded in 2018. Predating the founding of these two groups, mobilization for climate action had increased in 2015 following the high-profile Conference of the Parties (COP) to the United Nations Framework Convention on Climate Change (UNFCCC) held in Paris, where the post-Kyoto international climate regime was negotiated (Moor et al. 2021). Thus, overall, the degree of mobilization was high in 2018 and 2019.

However, for a complete assessment of the politicization of climate change, the third indicator, polarization, is paramount. To assess the degree of polarization, the analysis by Berker and Pollex (2023) is helpful since it offers insights into how the German political parties have reacted to FFF. The German party system comprises the Christian Democratic Union (CDU) and its Bavarian sister party, the Christian Social Union (CSU), and the Free Democratic Party (FDP) as the right-leaning parties, in particular on environmental issues. The left-leaning parties are the Social Democratic Party (SPD), the Greens, and the socialist Left. In addition, the German party system comprises the far-right populist party Alternative for Germany (AfD), which was first represented in the Federal Parliament in 2017. Berker and Pollex (2023) show that the Greens were the most supportive of FFF and the AfD the least supportive – a finding that resonates with the prevailing view in the literature, namely that right-wing populist parties tend to be dismissive of the anthropogenic origins of climate change (see, e.g., Lockwood 2018).

The AfD's stance on climate change aligns with the corresponding attitudes of the party's voters. Figure 2.2 presents survey data from the collaborative research project Digitalization in Dialogue. The data were collected in October 2021 and cover information on the attitudes and preferences of 1,075 respondents who, as

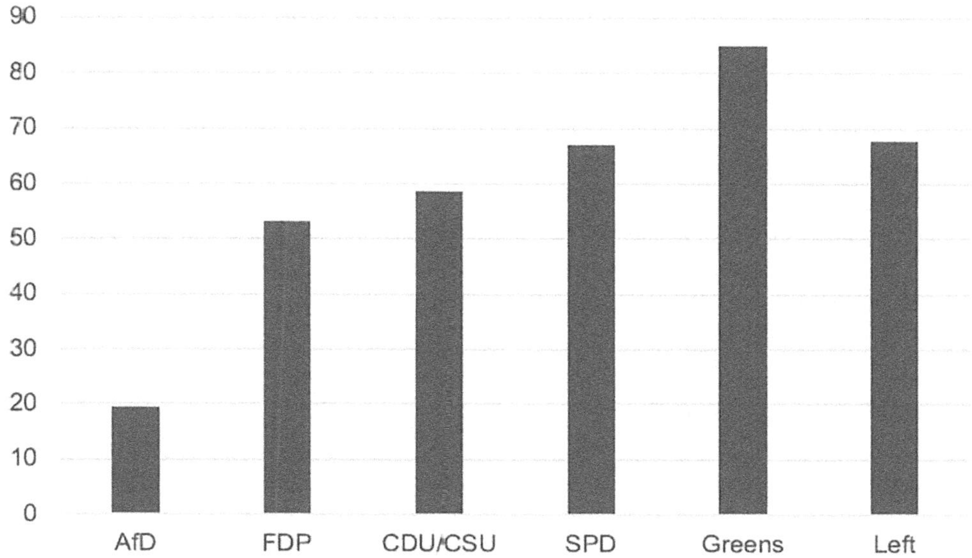

Figure 2.2 Percentage share of individuals indicating that climate change is mostly or only manmade, 2022.
Note: Own elaboration based on data from the Digitalization in Dialogue (https://digilog-bw.de/) collaborative project.

a whole, are representative of the German population in terms of age, gender, and education. The survey shows the share of survey participants who believe climate change is mostly or exclusively manmade. The responses are broken down by the political parties for which the participants intended to vote. The figure reveals that with almost 85 percent, the share of individuals agreeing that climate change is mostly or exclusively manmade is highest among supporters of the Greens and lowest among supporters of the AfD, where the agreement rate is 19 percent.

The AfD is the most vocal against climate action; while the other parties agree on the need for climate action, they differ in their preferred types of policy instruments or their levels of ambition. As the past electoral campaigns have witnessed, the mainstream parties agree on a set of issues, which results in a constellation in which they compete against each other, but jointly against the AfD. From this, the following situation emerges for polarization. In terms of whether climate action should be taken at all, the polarization concerns the AfD and its supporters on the one side and all the other mainstream parties and their supporters on the other. In relation to the level of ambition of climate policy, the Greens represent one extreme and the FDP the other. Altogether, the issue is polarized but not excessively so, since the main question for most parties and their supporters is not whether climate action is necessary but what kind of climate action they consider necessary or desirable.

Overall, the emergence of FFF helped to mobilize action for climate change and increased the salience of this issue among young people, which resulted in rising issue salience within the broader population. At the time when FFF began, the issue became polarized but not excessively so, since all of Germany's mainstream political parties agreed on the need for policy action, thereby distinguishing themselves from the AfD. Consequently, all three criteria for politicization as put forth by Wilde (2011) were met in Germany in the period between 2018 and 2021.

In line with Feindt, Schwindenhammer, and Tosun (2021), we hypothesize that this high, though not excessive, level of politicization resulted in an opportunity to adopt more ambitious policies addressing climate change. The impact of politicization on climate politics lends itself to more refined expectations because politicization has been widely applied in comparative politics already. Given the increased politicization of climate change vote-seeking, responsive political parties should not only integrate energy and climate policy issues into their election manifestos and basic policy programs but also develop a distinct policy position on them that resonates with the policy preferences of their (likely) voters. Furthermore, and given the support of a clear majority of German citizens and likely voters for increasing climate action (Bohdanowicz 2021), we postulate that all mainstream parties – CDU/CSU, SPD, Greens, and FDP – will converge in their positions on climate change over time, so that they promote more political activities to fight climate change. In line with its character as a right-wing populist party, the AfD should, conversely, adopt an opposite stance on climate policy compared to the center-right and center-left parties and should thus favor less state activity in fighting climate change, which is, according to the AfD 2016 party program, not caused by human activities. Because the socialist Left party also incorporates populist elements in their program, we expect them to be less in favor of more state activities to fight climate change, too.

2.2 Issue Politicization as a Driver of Policy Change

Because of its large industrial sector, Germany is one of the EU member states with above-average GHG emissions relative to its population. Between 2005 and 2019, the country's carbon emissions fell by 18 percent, indicating that some moderate-level climate action had been taken by the various governments of Chancellor Angela Merkel (2005–2021). However, Germany's performance lagged behind the EU-wide reduction of carbon emissions of 19 percent for the same period. The reduction in carbon emissions mostly resulted from an increase in the share of renewable energy, the switch from coal to gas, and the conversion of its coal-fired power plants into gas-fired ones, as well as from the implementation of the EU Emissions Trading System (European Parliamentary Research Service 2021).

The most challenging aspect of German climate policy has concerned the role of hard coal and lignite (Markard, Rinscheid, and Widdel 2021). The coal sector has been the backbone of the German industry and an important generator of jobs, so the regions that will be affected by the phaseout will need to undergo fundamental transformations and receive financial support in order to reduce the social impact. However, it is worth noting that the coal industry was also contested because of its adverse effects on the environment, but protests against it were mostly local. Compared to coal, nuclear power had been more contested in Germany, and the Greens have predominantly been anti-nuclear power since the formation of their West German predecessor in 1980. Therefore, nationwide environmental organizations had focused on mobilizing against nuclear power, paying limited attention to coal (Brauers, Oei, and Walk 2020).

However, the EU's announcement in 2017 that it would adopt stricter emission standards for large combustion plants (i.e. combustion installations with a rated thermal input exceeding 50 MW) burning coal and lignite as a fuel put pressure on the German federal government to take action. Therefore, even before FFF emerged, the ruling parties of the CDU/CSU and the SPD agreed in the coalition contract of March 2018 to phase out coal-fired power plants, which must be regarded as a radical shift from the previously existing policy regime.

To this end, the government launched, as stated in the coalition agreement, a "Commission on Growth, Structural Change and Employment" (known as the Coal Commission), which consisted of four chairs and twenty-four representatives from industry, unions, environmental NGOs, the German states, and selected scientists. Multistakeholder commissions are a typical feature of the German political system; as an instrument for incorporating expertise and interests from different groups, they have been informing policymaking in Germany since the 1960s (Müller-Hansen et al. 2022).

The Coal Commission was launched on June 6, 2018, and presented its policy recommendations on January 26, 2019. Most importantly, it recommended the gradual phaseout of coal-fired power generation, ending completely no later than the end of 2038. To achieve this goal, the proposal recommended that by 2022, the power generated from anthracite and lignite (two types of coal) should be reduced to around 15 GW. By 2030, this figure would be reduced further, to an output of about 8 GW for anthracite-fired power stations and 9 GW for lignite-fired power stations, respectively. By 2038 at the latest, the use of coal-fired power stations is to be completely ended (Commission on Growth, Structural Change and Employment 2019).

Some groups regarded the recommendations of the Coal Commission as a success, whereas scientists, climate experts, and social movements such as FFF criticized it for not being ambitious enough (Brauers, Oei, and Walk 2020). Analyzing the debate on the social media platform then known as Twitter, Müller-Hansen et al. (2022) show a negative trend of average sentiment scores throughout the entire period of study – from January 2017 to February 2020 – as well as a polarization of sentiments on the work of the Coal Commission over time. While the online debate on the coal phaseout was strongly affected by events related to the working of the Coal Commission, FFF and their online activities were also important for stimulating discussion among Twitter users. FFF held a negative view on both the recommendations of the Coal Commission and the Phaseout of Coal-Fired Power Plants Act adopted on July 3, 2020, which directly translated into law the recommendations made by the Coal Commission.

In addition to setting the wheels in motion for phasing out coal-fired power plants and banning the deployment of new coal-fired plants after August 14, 2020, with the exception of those that received their operation licenses before January 29, 2020, this act promised financial compensation to operators of coal-fired plants. This clause has been criticized by environmental and climate groups as being too generous (Gearino 2020) and reflects the Coal Commission's attention to the notion of a just transition and to transition assistance policies that target potential losers in the process (Bang, Rosendahl, and Böhringer 2022).

In addition to the Phaseout of Coal-Fired Power Plants Act, the federal government amended the German Renewable Energy Sources Act to codify the goal of raising the percentage of renewables to 65 percent by 2030. And through the Fuel Emissions Trading Act of January 2021, the federal government introduced a price of EUR 25 per ton of carbon dioxide emitted from the transport and heat generation sectors. The prices will increase in 2025, and from the following year onward emissions certificates will be auctioned (European Parliamentary Research Service 2021).

According to Brauers, Oei, and Walk (2020: 244), "rising civil society pressure, as well as from the coal regions demanding financial support, pushed the government to introduce a 'Commission on Growth, Structural Change and Employment' in 2018." While we agree with this statement, we contend that it is important to add that the German federal government was also under pressure to address the issue because of the EU. In fact, we are inclined to argue that it was above all the pressure from the EU that induced the Bundestag to make reforms and that it was the level of politicization, as it unfolded from the end of 2018 and throughout 2019, which affected the government's work on the Coal Commission and the level of ambition inherent in its recommendations.

Reflecting on the theoretical argument put forth by Feindt, Schwindenhammer, and Tosun (2021), we can state that politicization did not trigger policy change but facilitated and shaped the outcome of the corresponding process, while pressure exerted by FFF and other climate policy advocates resulted in a modification of the coal phaseout trajectory after the Greens became a junior partner in the coalition government of Olaf Scholz. In response to ongoing criticism that the phaseout would take too long, the Federal Parliament adopted on December 1, 2021, an act in order to bring forward to 2030 the phaseout of coal in the Rhenish lignite area in North Rhine-Westphalia, the most important coal-producing state.

2.3 Issue Politicization as a Driver of Electoral Competition

We now take a closer look at the positions the political parties' candidates adopted on climate change between 2017 and 2021, a period that covers the year 2018 when climate change became the most important problem in Germany as viewed by the public (see Figure 2.1). The focus on the candidate level allows for a more conservative test of our expectation that parties respond to the politicization of issues. The conservativeness of the approach results from the fact that candidates nominated by the respective political party normally follow the party line when positioning themselves on issues; however, they may choose not to, since the characteristics of the electoral districts in which the candidates compete for votes can set incentives to deviate from the party's position (Bäck and Debus 2019; Baumann, Debus, and Müller 2015). If the parties' candidates shifted their position on climate change between 2017 and 2021 in a decisive manner, then we can assume that the respective party changed its position not only at the leadership level but also at the candidate level.

A further advantage of focusing on the candidate level instead of the party leadership level is the availability of data that cover the positions of the parliamentary candidates on fighting climate change. We make use of the German Candidate Studies for the 2017 and 2021 Bundestag elections (GLES 2018, 2022). This dataset covers attitudes, policy preferences, and individual characteristics of the candidates for the 2017 and 2021 Federal Parliament elections and is part of both the German Longitudinal Election Study (GLES) and the Comparative Study of Electoral Systems (CSES) research network. The data were collected using a survey on the candidates of the six parliamentary party groups that were represented in the federal parliaments elected in 2017 and 2021. Participants of the candidate study were asked about their position on climate change by means of the following question: "Some believe that much more needs to be done in politics to combat climate change. Others believe that policies to combat climate change have already gone far too far. What is your opinion on combating climate change?"

All candidates were asked to participate in the study by mail or email; they included candidates from CDU and the CSU (the CSU only competes for votes in the state of Bavaria and is therefore considered in the following as a separate party), the SPD, the AfD, the FDP, the Left, and the Greens. In the case of the 2017 candidate study, 803 of the 2,516 candidates participated in the survey and completed it; in the 2021 candidate study, 735 partial and full interviews were carried out during the data collection process, which took place in the four months after the respective election.

Figures 2.3 and 2.4 show the distribution and median position of the candidates on climate change policy after the 2017 and 2021 elections. The candidates were asked about their position on a policy dimension differentiating between preferences for more climate action or less. In the case of the 2017 federal election, when migration and integration were the most important problems for the voters and climate change was not so salient among the population (see Figure 2.1), the candidates clearly differed – on average – in their positions on the issue of change. While candidates of the right-wing AfD and of the center-right parties the CDU, CSU, and FDP stated on average that existing climate policy was sufficient, left-leaning parties, and here in particular the Greens, called for more climate action (see Figure 2.3). There is,

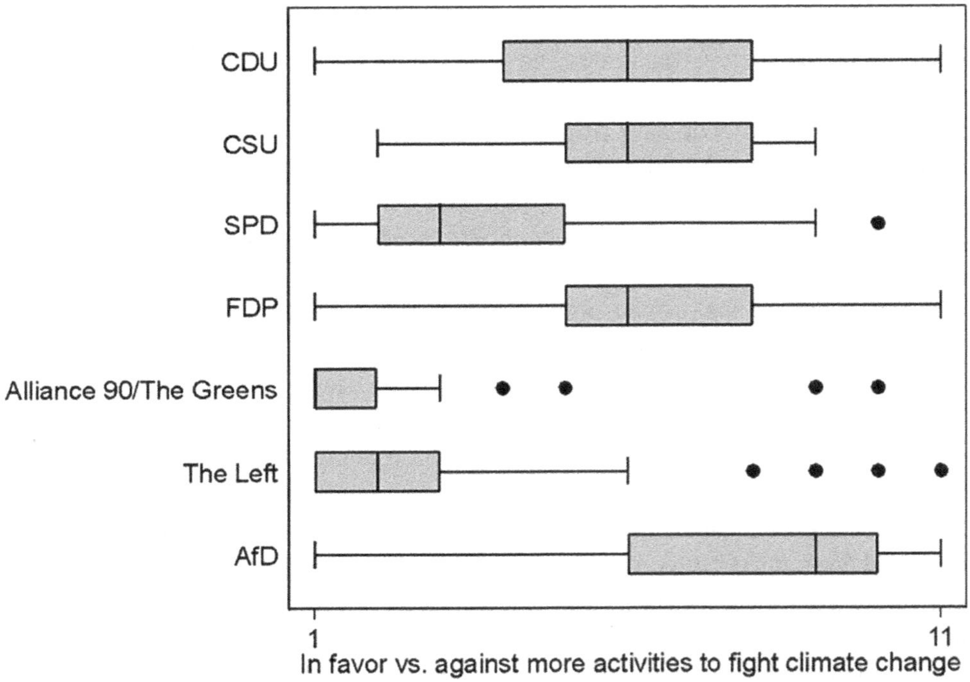

Figure 2.3 Climate policy positions of candidates of the major German parties for the 2017 Bundestag election.
Note: Own elaboration and calculation based on GLES (2018). The presented boxplots provide information on the distribution of the climate policy positions within the respective party. The line within the box shows the median position; the box provides information on the 25th and 75th percentile of the distribution of climate policy positions among the candidates of the parties. Dots indicate candidates that have climate policy positions outside the 1.5 interquartile range, which are indicates by the whiskers.

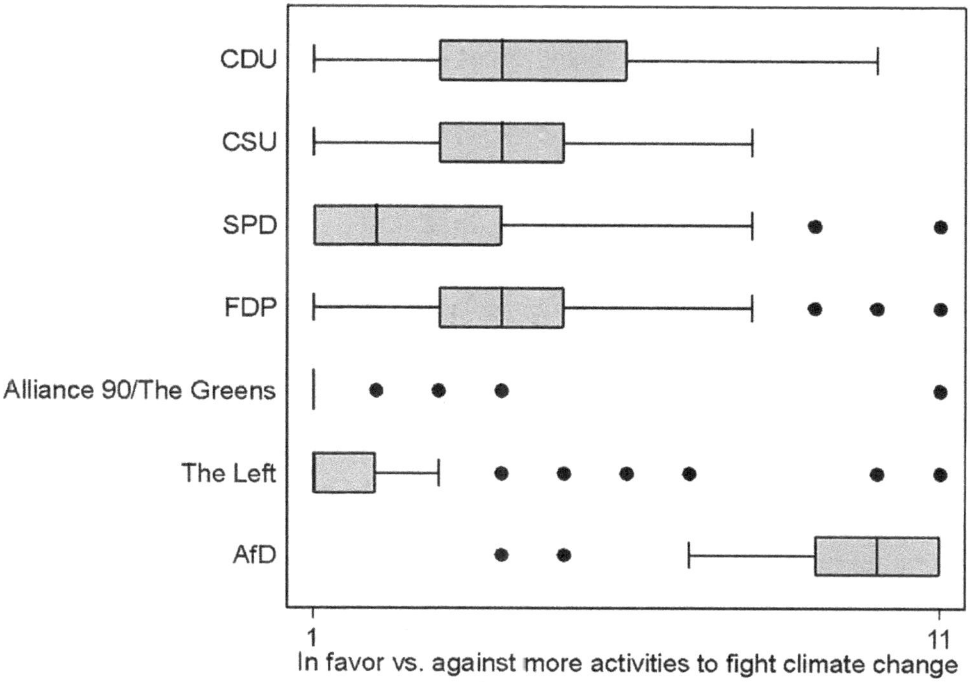

Figure 2.4 Climate policy positions of candidates of the major German parties for the 2021 Bundestag election.
Note: Own elaboration and calculation based on GLES (2022). See also the notes for Figure 2.3.

furthermore, a high level of intraparty diversity, in particular within the CDU and the FDP, indicating that the candidates of these parties varied a lot in their position on fighting climate change, possibly because their respective party leadership did not adopt a clear and consistent policy position on that very issue.

In 2021, when climate change became a politicized issue, the intraparty heterogeneity on fighting climate change decreased and the positions of the candidates converged clearly toward one that is more in favor of more ambitious climate action. The only exception regarding the latter is the AfD: The candidates of this party moved to the opposite side of this dimension and preferred, on average, significantly less climate action. In contrast to our expectation, the Left and its candidates did not change their position on climate change and preferred more climate action; however, there are, as the boxplot in Figure 2.4 indicates, a number of candidates of the Left who differed from this position and preferred less climate action. No such variation existed in the AfD. This indicates that the position of the candidates from the Left is less cohesive than it is for those of the AfD when it comes to climate action.

The positions of the two largest German parties – the "catch-all" parties the CDU/CSU and the SPD – were, in contrast to 2017, very similar on climate change in 2021, and even the market-liberal FDP showed a position similar to that of the CDU/CSU and the SPD with regard to support for climate action.

These findings not only align with our expectations but also support the reasoning of the studies by Kriesi and their coauthors (Kriesi et al. 2006, 2008), who regard environmental issues and related policy areas such as climate change as part of a broader societal cleavage that differentiates between left-libertarian parties, such as the Greens, on the one side, and right-wing populist parties, such as the AfD, on the other. These two parties form the extreme ends of the climate policy dimension we are studying here, which indicates that an increase in the salience and politicization of climate policy can result in greater support for the parties that formulate the clearest and most extreme positions on how to deal with climate change.

The convergence of the climate policy positions of the mainstream parties provides an opportunity for more ambitious climate action. This interpretation is supported, inter alia, by the fact that the 2022 act on the accelerated coal phaseout in the industrial Rhenish lignite area was supported by 523 Members of Parliament (MPs) and rejected by only 92, indicating cross-parliamentary support, including from factions affiliated with the opposition. Indeed, the oppositional Christian Democrats voted in favor of the bill initiated by the traffic light coalition (the FDP, the SPD, and the Greens), while the AfD and the Left voted against it (for details on this vote in the Bundestag, see Deutscher Bundestag 2022).

It will be interesting to see how this new political constellation acts in future climate policy proposals. What is already apparent is that the MPs of the different factions agree on the need to decarbonize the energy sector, though they hold differing views on the other sectors.

Most importantly, the decarbonization of the traffic sector is likely to spark more controversy than the measures targeting the energy sector. In fact, one of the main demands of FFF, as well as of more radical spin-off movements such as the Last Generation, is to introduce a general speed limit on the *Autobahn* (motorway), on which the views among Germans are heavily polarized. The polarization is also reflected in the parties' positions on this policy demand: The right-leaning parties – the CDU/CSU, the FDP, and the AfD – oppose a general speed limit, whereas the left-leaning parties support it. The introduction of general speed limit would mean a fundamental shift in Germany's identity as a "car state."

2.4 Discussion and Conclusion

In this chapter, we assessed the extent to which the emergence of FFF resulted in a politicization of climate change and how this affected climate policy and politics from 2018 to 2022. While the emergence of FFF itself is a manifestation of the issue of politicization, that is, of societal mobilization, it has been accompanied by an increase in the salience of climate change and a polarization of how political parties and their electoral candidates have positioned themselves on it. Thus, from 2018 onward, climate change underwent strong politicization; while this then decreased during the COVID-19 pandemic, it increased again afterward. In Germany, in late 2023, climate change was widely regarded as an important issue and one that requires policy action.

The politicization resulted in a situation in which the former federal government of Angela Merkel, after years of hesitation, took an important yet painful step to curb the country's carbon emissions: the gradual phaseout of coal-fired power plants. In light of the importance of industry and the coal sector to the country, this decision was of a transformative nature. A transformative policy approach is what Germany needs in order to reduce its carbon emissions to such a degree that they are proportionate to its population size and in line with the EU's reduction targets. However, the phaseout of coal-fired power plants also came at a price. The government granted the energy producers compensation payments which many criticized as being too high and benefiting loss-making power plants through the creation of windfall profits (Tiedemann and Müller-Hansen 2023).

As the analysis of the positions of the political parties and their electoral candidates has shown, the mainstream political parties have converged in their positions on climate change and the need for climate action. The only exceptions have and continue to be the AfD, which holds a dismissive stance on climate action, and some intraparty factions within the socialist Left. However, this convergence referred to climate policy in abstract terms and not to the specific policy measures supported by the individual parties. The current federal government is ideologically heterogeneous as it comprises two center-left parties – the SPD and the Greens – and the fiscally conservative FDP. Since they entered into office in December 2021, the policy positions of the Greens and the FDP in particular have diverged. In line with their ideology, the Greens have pushed for more ambitious climate action, which the FDP, also in line with its market-liberal ideology, has opposed. The best-known case is the FDP's opposition to the Building Energy law proposed by the federal minister of the economy, Robert Habeck (the Greens), to phase out gas and oil heating. The liberal FDP delayed the introduction of the bill to the Federal Parliament as it disapproved of its stipulations, which led to an increase in the number of intragovernmental conflicts.

The present situation – in which the Greens are a member of the federal government and actually propose climate policies while the issue is politicized – is different from the one before they entered government. Now the politicization rather results in parties diverging on the corresponding policy positions and in a more contentious climate politics. In the longer run, the politicization of the issue in combination with the stock of climate policies that entail not only benefits but also costs to certain groups may require strategies that ensure the stability of climate policies, for example by depoliticizing the issue. Otherwise, depending on the preferences of the electorate and of future governments, it might be strategically reasonable to dismantle climate policy.

Despite Russia's attack on Ukraine and the implications for Germany's energy mix, which used to rely heavily on Russian gas, the coal phaseout was not questioned while the shutdown of the last three operating nuclear power stations in April 2023 was. However, in November 2023 the federal minister of finance, Christian Lindner, of the FDP cast doubts on the country's ability to phase out coal-fired power plants by 2030. It remains to be seen how the politics around the coal phaseout will evolve given that in addition to Russia's war on Ukraine there now exists a conflict in the Middle East which could affect oil prices.

In their introduction to this volume, Paterson, Tobin, and VanDeveer state that the literature treats politicization and policy change as one strategy and depoliticization and policy stability as another. Based on the analysis of the German case between 2017 and 2021, we can confirm that politicization can change climate policy and politics. However, by drawing on the situation after 2021 especially, when more policy action took place thanks to the government participation of the Greens, the case can also offer a caveat that politicization cannot be the only mechanism for the long-term implementation of a transformative policy agenda. Politicization can entail an expansion in the density and the level of ambition of climate policy, but it can also make climate policy prone to dismantling and ultimately reduce the level of ambition. We know from the literature that dismantling is more difficult to achieve than expansion, but we equally know that it happens, even if less directly or less visibly (Burns and Tobin 2020).

We invite future research to pay more attention to the relationship between politicization, policy change, and changes in the political dynamics. What extensive literature otherwise seems to have neglected until now is how a changing stock of public policy adopted in response to politicization affects the degree of the politicization of the future political process. We suggest studying this aspect particularly for climate change governance, given that it will require sustained policy action for a long period of time.

References

Abou-Chadi, T. (2016). Niche party success and mainstream party policy shifts: How green and radical right parties differ in their impact. *British Journal of Political Science* 46(2), 417–436.

Bäck, H., and Debus, M. (2019). When do women speak? A comparative analysis of the role of gender in legislative debates. *Political Studies* 67(3), 576–596.

Bang, G., Rosendahl, K. E., and Böhringer, C. (2022). Balancing cost and justice concerns in the energy transition: Comparing coal phase-out policies in Germany and the UK. *Climate Policy* 22(8), 1000–1015.

Baumann, M., Debus, M., and Müller, J. (2015). Personal characteristics of MPs and legislative behavior in moral policymaking. *Legislative Studies Quarterly* 40(2), 179–210.

Berker, L. E., and Pollex, J. (2023). Explaining differences in party reactions to the *Fridays for Future*-movement: A qualitative comparative analysis (QCA) of parties in three European countries. *Environmental Politics* 32(5), 755–792.

Bohdanowicz, Z. (2021). Different countries, common support for climate change mitigation: The case of Germany and Poland. *Climate* 9(2), 27.

Brauers, H., Oei, P.-Y., and Walk, P. (2020). Comparing coal phase-out pathways: The United Kingdom's and Germany's diverging transitions. *Environmental Innovation and Societal Transitions* 37, 238–253.

Burns, C., and Tobin, P. (2020). Crisis, Climate Change and Comitology: Policy Dismantling Via the Backdoor? *Journal of Common Market Studies* 58(3), 527–544.

Carter, N. (2013). Greening the mainstream: Party politics and the environment. *Environmental Politics* 22(1), 73–94.

Commission on Growth, Structural Change and Employment. (2019). *Final Report*. Berlin: Federal Ministry for the Economy and Energy.

Deutscher Bundestag (2022). Braunkohlausstieg [Lignite phaseout]. Voting results. www.bundestag.de/parlament/plenum/abstimmung/abstimmung/?id=826

European Parliamentary Research Service. (2021). Climate action in Germany: Latest state of play. Briefing paper. www.europarl.europa.eu/RegData/etudes/BRIE/2021/690661/EPRS_BRI(2021)690661_EN.pdf

Farstad, F. M. (2018). What explains variation in parties' climate change salience? *Party Politics* 24(6), 698–707.

Feindt, P. H., Schwindenhammer, S., and Tosun, J. (2021). Politicization, depoliticization and policy change: A comparative theoretical perspective on agri-food policy. *Journal of Comparative Policy Analysis: Research and Practice* 23(5–6), 509–525.

Forschungsgruppe Wahlen E. V. (2024). Politik II [Politics II]: Long-term development of important trends from the political barometer on political issues. www.forschungsgruppe.de/Umfragen/Politbarometer/Langzeitentwicklung_-_Themen_im_Ueberblick/Politik_II/#Probl1

Gearino, D. (2020). What Germany can teach the US about quitting coal. *Inside Climate News*, October 15. https://insideclimatenews.org/news/15102020/germany-coal-transition/

GLES (German Longitudinal Election Study) (2018). *Kandidatenstudie (GLES 2017)*. Berlin: GLES.

GLES (German Longitudinal Election Study). (2022). *GLES Kandidierendenstudie 2021*. Berlin: GLES.

Grande, E., and Hutter, S. (2016). Beyond authority transfer: Explaining the politicization of Europe. *West European Politics* 39(1), 23–43.

Jahn, D., and Korolczuk, S. (2012). German exceptionalism: The end of nuclear energy in Germany! *Environmental Politics* 21(1), 159–164.

Kriesi, H., Grande, E., Lachat, R. et al. (2006). Globalization and the transformation of the national political space: Six European countries compared. *European Journal of Political Research* 45(6), 921–956.

Kriesi, H., Grande, E., Lachat, R. et al. (2008). *West European Politics in the Age of Globalization*. Cambridge: Cambridge University Press.

Lockwood, M. (2018). Right-wing populism and the climate change agenda: Exploring the linkages. *Environmental Politics* 27(4), 712–732.

Markard, J., Rinscheid, A., and Widdel, L. (2021). Analyzing transitions through the lens of discourse networks: Coal phase-out in Germany. *Environmental Innovation and Societal Transitions* 40, 315–331.

Moor, J. D., de Vydt, M. D., Uba, K., and Wahlström, M. (2021). New kids on the block: Taking stock of the recent cycle of climate activism. *Social Movement Studies* 20(5), 619–625.

Müller-Hansen, F., Lee, Y. T., Callaghan, M., Jankin, S., and Minx. J. C. (2022). The German coal debate on Twitter: Reactions to a corporate policy process. *Energy Policy* 169(1), 113178.

Tiedemann, S., and Müller-Hansen, F. (2023). Auctions to phase out coal power: Lessons learned from Germany. *Energy Policy* 174, 113387.

Tosun, J., Geese, L., and Lorenzoni, I. (2023). For young and future generations? Insights from the web profiles of European Climate Pact Ambassadors. *European Journal of Risk Regulation* 14(4), 747–759.

Venghaus, S., Henseleit, M., and Belka, M. (2022). The impact of climate change awareness on behavioral changes in Germany: Changing minds or changing behavior? *Energy, Sustainability and Society* 12(1), 121.

Wilde, P. D. (2011). No polity for old politics? A framework for analyzing the politicization of European integration. *Journal of European Integration* 33(5), 559–575.

Wilde, P. D., and Zürn, M. (2012). Can the politicization of European integration be reversed? *JCMS: Journal of Common Market Studies* 50, 137–153.

3

Climate Change Worldviews and the Scale of Environmental Justice

MICHAEL MÉNDEZ

California has provided a test case over the last twenty years, showing other governments how disparate interest groups – environmental justice activists, lobbyists, industry, labor unions, policymakers, and scientists – can collaborate and stabilize comprehensive climate action policies (Méndez 2020; Rabe 2009).[1] Such negotiations can be contentious, in part because participants' ideas of what climate change solutions should entail are often shaped by unspoken worldviews (the social, political, and cultural attitudes toward the world that orient people's actions). These differences in worldview, if unacknowledged, can lead to the breakdown of trust among groups that are nominally working toward the same goal: reducing the harm that climate change will do to human societies and our planet (Ballew et al. 2019: 3; Hoffman 2015). Nonetheless, confrontations and collaborations between groups with differing worldviews have the potential to reshape environmental governance and power relations in society. During my senior policy roles in California (2003–2023), I witnessed and analyzed the effects of incompatible worldviews, as they played out in the conduct of climate change politics.[2] This chapter outlines the two most salient worldviews, arguing that the gulf between them could only be bridged when action was *repoliticized* to include multiple scales, ensuring climate change solutions tackle both the global problem and local needs.[3]

According to the political scientist Michael Lind (2011),

> a worldview is a coherent understanding of the nature of reality, which permits its holders to interpret new information in light of their preconceptions. Clashes among worldviews cannot be ended by a simple appeal to facts. Even if rival sides agree on the facts, people may disagree on conclusions because of their different premises.

[1] The US Environmental Protection Agency (2012) has conceptually defined "environmental justice" as the "fair treatment and meaningful involvement of all people regardless of race, color, national origin, or income with respect to the development, implementation, and enforcement of environmental laws, regulations, and policies."

[2] Méndez previously served in California as a senior consultant, lobbyist, and gubernatorial appointee during the passage of the state's internationally acclaimed climate change legislation.

[3] Material in this chapter has been adapted from Méndez (2020).

Lind argues that this is why politicians often seem to speak past one another or ascribe different meanings to the same events. In this sense, a worldview can include ideas about nature, values, emotions, and ethics (Ballew et al. 2019). Climate change worldviews can be placed on a spectrum between two opposing positions, which I refer to as "carbon reductionism" and "climate change from the streets." What follows is a generalization of the dynamics I observed during the initial years (2006–2012) of implementation of California's climate change policies. While the perspectives of particular people and organizations were always more complex than these categories may suggest, they are nonetheless useful in developing a sense of the sources of misunderstanding, disagreement, and fracture that can hamper public discussion of such policies, as well as some of the strategies that allow differing worldviews to be brought into creative tension and stabilized (Hulme 2007, 2010).

The case of California is particularly productive in this regard. As the world's fourth-largest economy and the only US state to implement a comprehensive program of regulatory and market-based mechanisms to reduce greenhouse gas emissions, California has consistently been at the forefront of broader national and global environmental experimentation (Stone 2012). The state's cap-and-trade program – a central, market-based mechanism for ensuring emissions reductions – is the third largest in the world, after those of the European Union and China (CARB 2018). This program has been especially contentious in debates within California; supporters emphasize its global reach and cost-effectiveness, and detractors criticize its inequitable effects on specific local communities and demographic groups. California's prominence in climate change policy makes it an ideal place to investigate the dynamics of such disputes, and their roots in differing climate change worldviews (London et al. 2013).

What I refer to as carbon reductionism was broadly based on utilitarianism – efforts to develop climate change policy that would deliver the "greatest good for the greatest number." Utilitarianism is an ethical theory which states that the best action is the one that maximizes utility for the greatest number of people in society (Bentham 2007). This worldview tends to be associated with regulators, traditional environmentalists, climate scientists, multinational corporations, and global bodies such as the United Nations. Carbon reductionism reflects an adherence to cost-effectiveness and market-based solutions (i.e. cap-and-trade), focused on reducing global greenhouse gas emissions. It is argued that although emissions of greenhouse gases and local pollutants may be correlated, greenhouse gas mitigation efforts should not be required to reduce local pollution. These are seen as fundamentally different problems that are best addressed separately. "Carbon reductionism" thus literally refers to stakeholders' desire to reduce carbon dioxide emissions (the most prevalent greenhouse gas) while simultaneously emphasizing the reductive logic embedded in this approach. Under this perspective, climate change policies are judged on whether they are economically advantageous and benefit the majority. While traditional environmentalists use a market logic in hopes of creating climate policy that can be replicated elsewhere, many businesses are in favor of a limited focus because it minimizes government intervention and supports risk management and corporate strategies that lessen the financial burden of mitigation efforts (Cushing et al. 2018). In this respect, carbon reductionism is similar to the pursuit of "stability" in the climate policymaking process, which often seeks the durability of programs, ultimately to the benefit of powerful interest groups.

Environmental justice groups, in contrast, generally emphasize moral rights, which leads them to a critical reevaluation of both the practice and the politics of reducing carbon emissions. One result is a refusal to uncouple the global effects of greenhouse gases from the local health effects of other pollutants that are emitted with them. The environmental justice movement originated in civil rights campaigns that worked to expose the socially uneven impacts of industrialization on low-income communities of color, and its focus remains on race, class, and the distribution of environmental hazards (London et al. 2013). Shaped by this history of activism, the worldview I refer to as "climate change from the streets" prioritizes equity and justice; from this perspective, the utilitarian approach often ignores distinctions between people and the disproportionate impacts of climate change on low-income communities of color. Climate change solutions are evaluated on their ability to address environmental disparities and prioritize communities living near polluting sources (Bullard 2000; Méndez 2020).

Environmental justice groups that support climate change from the streets legitimize knowledge based on lived experience within their communities and promote participatory, embodied, and experimental methods in the development of climate change policy. In contrast to carbon reductionism's focus on incremental greenhouse gas reductions within an existing economic framework, they are more willing to consider aggressive policies to reduce emissions and transition away from a fossil fuel economy. In climate change policy debates, environmental justice advocates are apprehensive about market-based solutions because they see them as serving those with wealth and power, rather than the disadvantaged (Kaswan 2018: 492; Schlosberg and Collins 2014: 364). For them, the main threat (their risk perception) from climate change is the disproportionate harm that it causes to health in their communities, including increased illnesses, injuries, and deaths from extreme events such as hurricanes and wildfires, or respiratory illnesses caused by degraded air quality that is further worsened by prolonged droughts and wildfires (Méndez 2020). This insight leads them to view climate change as an embodied phenomenon that affects people's daily lives in multiple ways. Therefore, the main argument of activists is that environmental protection and improving human health are inextricably linked, and maintaining that link is key to advancing future climate action policies (Krieger 2005). In essence, activists seek the *repoliticization* of climate policymaking to bring processes back into explicitly political arenas, as a method to expose existing power relations, and to make conflicts of power and interest apparent.

In what follows, I describe the general characteristics that distinguish carbon reductionism and climate change from the streets, pointing out the sources of tension between the two. Through this synthesis, we can see that responses to climate change are components of larger social, political, and environmental dynamics that combine to shape individuals' ideas about fairness and justice, and the role the government should play in enacting solutions. Understanding this larger picture, and examining often unspoken assumptions, allows us to reflect on how common ground can be deliberately negotiated between these positions – in particular, how environmental justice and public health can meaningfully be integrated into climate change policy. Practices that bridge between the scales that these two worldviews prioritize – the global and the local – can generate solutions that are truly innovative and work toward both climate change goals and social equity.

3.1 Carbon Reductionism

As public concern over the changing climate grows, governments and scientists are focusing more on its chief cause: global greenhouse gas emissions. This focus results in climate change policy with the specific goal of reducing seven emissions, rather than parallel policy goals. The seven gases have been identified by the Kyoto Protocol and California's Global Warming Solutions Act of 2006 – the state's main enabling legislation for its suite of regulatory programs (CARB 2008; IPCC 2014). Greenhouse gas emissions have "no direct public health impacts" since they are global pollutants that mix uniformly in the atmosphere; they do not have localized effects like particulate matter and nitrogen oxides (CARB 2008). Evidence of observed climate change impacts, moreover, is reported as strongest and most comprehensive for natural systems, a finding that is often used to justify a focus on biophysical rather than social systems (IPCC 2014).

Greenhouse gases and local air pollutants such as particulate matter and nitrogen oxides (precursors of smog) are often emitted concurrently from processes such as fossil fuel burning. But within the carbon reductionist framework, the most scientifically rigorous and cost-effective method to address climate change is understood to require a strict delineation between global and local scales (Cushing et al. 2018). Carbon dioxide (CO_2) holds a privileged position, since it is the most abundant anthropogenic greenhouse gas that contributes to global warming and persists in the atmosphere for many years. To quantify and monitor greenhouse gases, climate analysts convert the gas levels to a "CO_2 equivalent." The CO_2 equivalency of any given gas is calculated by multiplying its mass by the "global warming potential," which indicates the equivalent greenhouse effect of a pound, or metric ton, of the gas as compared to a pound, or metric ton, of CO_2 (CARB 2008).

Calculations centered on CO_2 equivalency and global warming potential are often linked with a view of nature that focuses on ideas such as biodiversity, ecological integrity, and natural systems. The ambitions of this type of climate change governance are often stated in quantitative terms, such as achieving carbon reduction targets, preventing dangerous anthropogenic interference with the climate system, and limiting the average global surface temperature increase. Given the assumption that climate change should be understood and addressed at the global scale, these ideas provide crucial tools for detecting its progress, measuring its causes and effects, and quantifying the changes necessary to human behavior to avoid catastrophe for our species and planet. Nonetheless, if used exclusively in public discourse, they can create an abstraction of nature and limit other types of knowledge about the changing climate, including environmental justice perspectives (Hull 2006: 4–5). These conceptual structures, moreover, narrow the focus of climate change measures to reducing greenhouse gas emissions, regardless of place or scale. Since climate change is a global issue, and greenhouse gases are global pollutants, it is argued, specific locations for reducing emissions do not matter, as long as reduction targets are achieved. However, if greenhouse gas reductions are coupled with reductions in other associated pollutants, location becomes significant for communities near major polluting sources like oil refineries. As these associated pollutants have direct and immediate health effects, focusing on the distribution of health benefits from local air quality improvements could increase public support for climate change policy (Cushing et al. 2018).

Second, carbon reductionism requires that climate change policy be supported by a community of scientific experts, whose knowledge is considered to be definitive. Under such an approach, dispassionate experts advise policymakers of objective truths, after which policymakers factor in social or political considerations. This process implies a linear approach in which scientific facts inform policy, and scientific inquiry is understood to be independent of society, politics, and values. This separation of science from politics results in policymaking that is seen as evidence-based and rational but can also be perceived as exclusionary, since only experts' knowledge and perspectives are valued (Ezrahi 1990; Jasanoff 2005). For example, when the scoping plan (a proposed framework for achieving greenhouse gas reduction targets) was adopted in 2008, the entire membership of the California Air Resources Board and its executive staff was white – and presumably did not share similar histories of environmental racism as activists. By 2011, when cap-and-trade was approved, the board's racial composition included one African American board member and one Asian American on its executive staff. Through the Global Warming Solutions Act's implementation, we see how leadership in climate change policy can be highly homogeneous in terms of race, gender, and class.

The third characteristic of carbon reductionism is the assertion that appropriate action on climate change requires an array of measures to capture the maximum technologically feasible and "cost-effective" emissions reduction opportunities wherever possible. The Global Warming Solutions Act defines the measure of cost-effectiveness as the "cost per unit of reduced emissions of greenhouse gases adjusted for its global warming potential" (California Health and Safety Code, §38505(d), §38560, §38561). The concept of cost-effectiveness, as applied to the reduction of carbon emissions, requires policymakers to judge the distribution of costs associated with implementing a climate change policy and account for any trade-offs. Attaining cost-effectiveness, however, does not fully allow the specification of where emissions abatement will take place. This raises broad questions: How do different climate policies affect the distribution of costs, benefits, and consequences? And how are these effects experienced across scales and demographic groups? Therefore, what seems like cost-effective policy from a carbon reductionist standpoint could appear to reproduce long-term injustices when viewed from the streets, since it might ignore, reinforce, or potentially worsen existing environmental and health disparities (Parks 2009).

Fourth, cost-effectiveness is often achieved through market-based mechanisms (such as cap-and-trade). This approach is utilized for large industrial emitters, such as electrical generation, manufacturing, cement production, and oil and gas production and supply. According to carbon reductionist logic, incentivizing mitigation and allowing regulated entities flexibility in deciding how and where best to meet reductions targets spurs market innovation and drives new technologies to higher volumes and lower prices (CARB 2014: 104). The cap-and-trade program accounts for less than one-third of the state's total mitigation measures, but it remains a central concern for environmental justice groups. The industrial sector's obligations to greenhouse gas reductions are achieved mainly through its compliance under the program, rather than reducing emissions directly, which activists argue would improve local air quality around polluting industrial facilities (Nachbaur et al. 2012).

The fifth characteristic of carbon reductionism is the geographic neutrality of its policy interventions. In shaping California's Global Warming Solutions Act, policymakers narrowed climate change measures to directly address greenhouse gas reductions across polluting sources, regardless of place or scale. State policymakers view California as a member of the global community and envision its policies to be part of a larger domestic and international system. The state has strategically

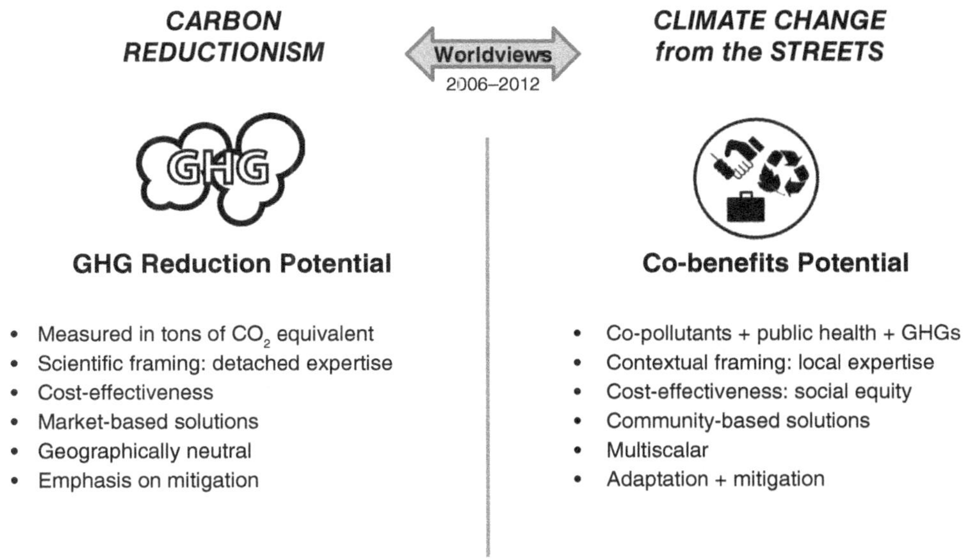

Figure 3.1 Tension between California's climate change worldviews.

chosen to move away from a "parochial" or neighborhood-level scale – even if attention to local conditions would result in the direct reduction of associated, harmful pollutants (CARB 2008).

The last defining characteristic of carbon reductionism is its main focus on mitigation. California's mitigation measures are intended to slow the emissions rate for human-caused greenhouse gases to avoid further disruptions to the earth's atmosphere. As adopted in 2006, the Global Warming Solutions Act did not initially include local adaptation measures. The goal of adaptation measures is to protect lives, health, property, and ecosystems from actual or anticipated climate change impacts, such as heat waves, droughts, wildfires, and flooding. Put another way, mitigation can be viewed as activities that protect "nature from society," whereas adaptation involves ways of "protecting society from nature" (ICELI 2009; Park 2009). Although robust adaptation strategies were developed in later years, early strategies were directed toward protecting hard assets such as vital infrastructure or ecological systems – not neighborhoods and socially vulnerable populations (Figure 3.1; Méndez 2015, 2016).

3.2 Climate Change from the Streets

Carbon reductionism may be the dominant climate change policy paradigm, but California environmental justice advocates are pushing forward an alternative. "Climate change from the streets" is a form of repoliticization that challenges the established policy practices of carbon reductionism by embedding justice and public health goals within climate change science and solutions. It shifts the focus of policies from the degradation of "nature" to address the simultaneous degradation of communities and asymmetries in the policymaking process. Environmental justice advocates know and analyze the effects of climate change through people's histories, culture, and embodied knowledge rather than solely through data gathered by experts and policy implemented by regulatory agencies (Park 2009).

In an effort to challenge the prevalence of carbon reductionism in climate change policy and action, environmental justice advocates have developed a community-based framework for addressing local and global environmental health impacts. Through this framework, they are demanding a greater role in the decision-making processes that impact their lives. They are debating not only the relationship between political power and climate expertise but also the efficacy of solutions to produce positive social change (Corburn 2005). This approach acknowledges how climate change is connected with other types of knowledge about the environment and enables different ways of knowing to play a valid part in framing policy responses. In climate change from the streets, individuals and groups with a range of competing interests work together to develop climate change policy and enact environmental justice (Méndez 2020).

The first characteristic that I attribute to climate change from the streets is that it emphasizes the need for climate change policies that yield substantial health benefits through the reduction of other pollutants that are commonly emitted alongside carbon dioxide. Environmental justice groups reference valuation studies that suggest that the value of such health benefits may be comparable in magnitude to that of reduced carbon emissions. They recommend that policymakers directly compare the cost of climate change actions with the economic value of their benefits, in terms of avoided damage to human health and the environment. For this reason, activists argue that cost-effective and efficient policy design should seek greater emissions reductions where health benefits would be highest and are most needed (Méndez 2020). Climate change policy that ignores such benefits is considered inefficient in two ways: It would choose suboptimal emissions reductions targets overall and it would fail to account for differences in abatement benefits across emission sources (Boyce and Pastor 2013: 803; Cushing et al. 2018; Méndez 2020).

Second, environmental justice groups center their work on issues of embodiment and ask, "What are the connections that their bodies make and manifest daily between the changing climate, pollution, and health?" (Krieger 2005: 2). This entails a holistic understanding of the multiple harms that pollution and a changing climate have on human bodies in specific local settings. These types of questions and forms of embodied knowledge, when engaged in truly collaborative settings, are a crucial supplement to solutions developed by policy experts. They confront uneven power dynamics in environmental governance and can alleviate problems that can be caused by introducing one-size-fits-all policy into various local contexts. Environmental justice groups, moreover, debate with experts over issues of truth and method, specifically challenging the political use and control of expertise by claiming to speak credibly as experts in their own right (Corburn 2005).

A third point is that environmental justice groups conceptualize climate change policy in a multifaceted way, calling out unequal impacts while advocating solutions focused on social and health equity. Health equity is achieved when every person has the opportunity to "attain his or her full health potential" and no one is "disadvantaged from achieving this potential because of social position or other socially determined circumstances" (Braverman et al. 2003: 181). In the context of climate change, this approach involves exploring how programs, practices, and policies affect the health of individuals, families, and communities. It establishes common goals and ongoing constructive relationships between the health sector, climate science, urban planning, and other fields at multiple scales (London et al. 2013).

Fourth, in promoting climate action from the streets, environmental justice groups seek to empower individuals and groups with the skills they need to effect change within their communities. They advocate for partnerships with residents, governments, and other entities as a means to harness policy processes that support community-defined goals. Hence, their solutions to climate change involve diverse stakeholders in the strategic and management activities of climate planning and policy (Corburn 2005). These approaches, such as transit-oriented development, renewable energy, and urban forestry projects, are aimed at reducing global greenhouse gas emissions and the risk of asthma and respiratory diseases. Advocates also seek to generate career-track jobs in the green economy for workers from disadvantaged communities. These solutions seek to target policy investments and resources to the neighborhoods most in need.

The fifth defining characteristic of climate change from the streets is that its practitioners make concerted efforts toward neighborhood-scale adaptation planning, in addition to mitigation measures. Such integration can be difficult because of important differences in policy objectives. Mitigation deals with the causes of climate change (accumulation of greenhouse gases in the atmosphere), whereas adaptation seeks to prepare society for the impacts of a changing climate. The policies are also defined in different spatial and temporal scales. For example, the benefits of adaptation are local and short-term, whereas mitigation benefits are global and long-term (Locatelli 2016: 1). Nonetheless, environmental justice groups contend that synergies between mitigation and adaptation can be developed in a cost-effective and equitable way. This type of policy integration is important because it acknowledges that certain irreversible and significant impacts from climate change are already underway and are inevitable, even if governments succeed in reducing greenhouse gas emissions. Environmental justice advocates argue that some groups are more socially vulnerable and will need additional safeguards from the immediate and anticipated effects of climate change (Stehr and Storch 2005).

Finally, climate change from the streets takes environmental justice advocates, whose concerns are often rooted in local conditions, beyond a single site and links them with the global scale of climate change, traversing local, regional, statewide, national, and international settings. How to take interconnected action across scales has become a central concern for these groups. Doing so creates opportunities (through repoliticization processes) to rethink the relationships among places, people, projects, and sources of knowledge, and opens up spaces that are often rendered invisible from the point of view of a single location or scale. The source of this challenge, and this potential, lies in the nature of a movement with local roots seeking to address a global crisis. Strategies for bridging between scales are crucial for bringing opposing climate change worldviews into dialogue and stabilizing climate-related regulatory program.

3.3 Conclusion

Through an analysis of climate change worldviews, we can better understand the culturally contingent nature of climate policy – the assumptions and values that often create conflict between community understandings of local environmental conditions and the prevailing top-down regulatory culture of climate change (Knox-Hayes 2016). In California, tensions between worldviews, moreover, are often centered on the politics of scale, markets, race, and class. The role of scale, in particular, can be difficult to grasp, since its subjective and political nature often goes unrecognized. Seeing a problem at any given scale involves decisions – conscious or not – about which of its aspects to disregard, and which to act upon (Williamson 2015: 19).[4] Scale is not objective. It is constructed through human relationships and is an important factor in political strategies and contests over power and authority. Changing the scale at which a problem is addressed can alter the power relationships that surround it – associations that determine unequal access to resources and institutions, and the ability to choose and act despite resistance from others (Osofsky 2009).

[4] According to Hari Osofsky (2009: 130–133), referencing climate change as a multiscalar problem can be a complex and contested concept in both geography and ecology. Geographers define scale through four aspects: (1) "a nested hierarchy of bounded spaces of differing size"; (2) "the level of geographic resolution at which a given phenomenon is thought of, acted on or studied"; (3) "the geographical organizer and expression of collective social action"; and (4) "the geographical resolution of contradictory processes of competition and cooperation" (Brenner 2004: 9). Ecologists, on the other hand, define scale in more technical terms as comprised of two parts: grain and extent. Grain refers to "the finest level of spatial or temporal resolution available within a given data set" and extent refers to "the size of the study area or the duration of the study" (Sayre 2005: 281).

Any account of climate change policy that focuses on a single scale thus can only be partial and undermine the stability of climate change programs. It is undeniable that no single locality, acting alone, can hope to address the problem. But analysis at the global scale will inevitably gloss over the question of how public definitions of climate change can reinforce existing local inequalities in power, resources, and health. If climate change is seen solely as a global phenomenon, then it will seem self-evident that only "global actors" – national governments, the United Nations, multinational corporations, the international community of scientists – are empowered to address it. By seeking to understand the effects of climate change and climate policy at multiple scales (and through the processes of repoliticization), we can promote more egalitarian forms of public decision-making about this critical issue (Barnett 2013).

Activists' advocacy work shows how climate change policy is an ongoing cultural creation, made and remade through the daily practices of diverse people. Through a reoccurring process of conflict and collaboration, a broad range of individuals and organizations are co-constituting what climate change means. The geographer Mike Hulme (2007) argues that the tension between worldviews can have a balancing – even creative – impact, yielding stronger, more robust approaches to resolving climate change. Furthermore, worldviews are not fixed and can transform over time. Scientific ideas and beliefs about climate change evolve together with the representations, identities, debates, and institutions that give practical effect and meaning to policies. In other words, the ways in which we conceptualize climate change don't just happen. People are behind our government, policies, and environmental values – and they can change their minds (Jasanoff 2005).

Beginning in 2012, California has moved away from a strict adherence to carbon reductionism by adopting a wave of legislation inserting and stabilizing environmental justice and public health elements into the state's climate change policies. The California Air Resources Board (charged with overseeing the state's climate programs) now has an environmental justice-focused senior officer and two board positions dedicated to environmental justice representatives. The state expanded its climate adaptation strategies to include robust local programs in disadvantaged communities. In addition, due to activists' campaigns, more than 50 percent of the state's cap-and-trade revenue must benefit disadvantaged communities – this has resulted in billions of dollars in investments (CARB 2018).

These successes have been underpinned in part by repoliticization processes at the city scale, where open-ended consultation and collaboration between activists, regulators, and elected officials brought different worldviews – and concerns of varying scales – into dialogue (Méndez 2015, 2020). They also reflect the innovative networked strategies that marginalized communities in California and beyond have used to narrate the effects of international carbon trading at a global scale, without reducing their local experiences to a single story (London et al. 2013). The science of climate change may be certain, but policy decisions about how to respond to its effects engage with complex social, cultural, and political realities. Understanding the power relations and worldviews of interested parties in this new policy arena must be a central focus of theory and practice as climate change laws, policies, and programs continue to develop.

References

Ballew, M., Goldberg, M., Rosenthal, S., Cutler, M., and Leiserowitz, A. (2019). Climate change activism among Latino and White Americans. *Frontiers in Communication* 3(58), 1–15.

Barrett, A. (2013). The necessity of a multiscalar analysis of climate justice. *Progress in Human Geography* 37(2), 215–33. https://doi.org/10.1177/0309132512448270

Bentham, J. (2007). *An Introduction to the Principles of Morals and Legislation*. Mineola, NY: Dover Publications.

Boyce, J. K., and Pastor. M. (2013). Clearing the air: Incorporating air quality and environmental justice into climate policy. *Climatic Change* 120(4), 801–14. https://doi.org/10.1007/s10584-013-0832-2

Braveman, P. A. (2003). Monitoring equity in health and healthcare: A conceptual framework. *Journal of Health, Population, and Nutrition* 21(3), 181.

Brenner, N. (2004). *New State Spaces: Urban Governance and the Rescaling of Statehood*. Oxford: Oxford University Press.

Bullard, R. D. (2000). *Dumping in Dixie: Race, Class, and Environmental Quality* (3rd ed.). Boulder, CO: Routledge.

CARB (California Air Resources Board). (2008). AB 32 Scoping Plan. https://arb.ca.gov/cc/scopingplan/scopingplan.htm

CARB (California Air Resources Board). (2014). AB 32 Scoping Plan Update. https://ww2.arb.ca.gov/our-work/programs/ab-32-climate-change-scoping-plan

CARB (California Air Resources Board). (2018). California Climate Investments Using Cap-and-Trade Auction Proceeds. www.arb.ca.gov/cc/capandtrade/auctionproceeds/2018_cci_annual_report.pdf

Corburn, J. (2005). Street science: Community knowledge and environmental health justice. In *Urban and Industrial Environments*. Cambridge, MA: MIT Press.

Cushing, L. J., Blaustein-Rejto, D., Wander, M. et al. (2018). Carbon trading, co-pollutants, and environmental equity: Evidence from California's cap-and-trade program (2011–2015). *PLoS One* 15(7), 1–20.

Ezrahi, Y. (1990). *The Descent of Icarus: Science and the Transformation of Contemporary Democracy* (1st ed.). Cambridge, MA: Harvard University Press.

Hoffman, A. (2015). *How Culture Shapes the Climate Change Debate*. Stanford, CA: Stanford University Press.

Hull, R. B. (2006). *Infinite Nature*. Chicago: University of Chicago Press.

Hulme, M. (2007). Geographical work at the boundaries of climate change. *Transactions of the Institute of British Geographers* 33(1), 1–11.

Hulme, M. (2010). Problems with making and governing global kinds of knowledge. *Global Environmental Change* 20(4), 558–64. https://doi.org/10.1016/j.gloenvcha.2010.07.005

ICELI. (2009). *The Mitigation-Adaptation Connection: Milestones, Synergies and Contradictions*. ICLEI – Local Governments for Sustainability. http://itepsrv1.itep.nau.edu/itep_course_downloads/~GeneralAQInfo/Climate%20Change/ICLEIMitigationAdaptationConnection.pdf

IPCC (Intergovernmental Panel on Climate Change). (2014). *Climate Change 2014 Synthesis Report Summary for Policymakers*. IPCC. www.ipcc.ch/pdf/assessment-report/ar5/syr/AR5_SYR_FINAL_SPM.pdf

Jasanoff, S. (2005). *Designs on Nature: Science and Democracy in Europe and the United States*. Princeton, NJ: Princeton University Press. http://grail.eblib.com.au/patron/FullRecord.aspx?p=713600

Kaswan, A. (2018). A broader vision for climate policy: Lessons from California. *San Diego Journal of Climate and Energy Law* 9, 83–150.

Knox-Hayes, J. (2016). *The Cultures of Markets: The Political Economy of Climate Governance*. Oxford: Oxford University Press.

Krieger, N. (2005). *Embodying Inequality: Epidemiologic Perspectives*. Abingdon: Routledge.

Lind, M. (2011). Michael Lind: The five worldviews that define American politics. *History News Network*, November 1. https://historynewsnetwork.org/article/135314

Locatelli, B., Fedele, G., Fayolle, A., and Baglee, A. (2016). Synergies between adaptation and mitigation in climate change finance. *International Journal of Climate Change Strategies and Management* 8(1). https://doi.org/10.1108/IJCCSM-07-2014-0088

London, J., Karner, A., Sze, J. et al. (2013). Racing climate change: Collaboration and conflict in California's global climate change policy arena. *Global Environmental Change* 23(4), 791–99. https://doi.org/10.1016/j.gloenvcha.2013.03.001

Méndez, M. (2015). Assessing local climate action plans for public health co-benefits in environmental justice communities. *Local Environment* 20(6), 637–63. https://doi.org/10.1080/13549839.2015.1038227

Méndez, M. (2016). From the street: Civic epistemologies of urban climate change. In D. Wendel and F. Samuels, eds., *Spatializing Politics: Essays on Power and Place*. Cambridge, MA: Harvard University Graduate School of Design, pp. 337–66.

Méndez, M. (2020). *Climate Change from the Streets: How Conflict and Collaboration Strengthen the Environmental Justice Movement*. New Haven, CT: Yale University Press.

Nachbaur, J., and Roberts, T. (2012). *Evaluating the Policy Trade-Offs in ARB's Cap-and-Trade Program*. Sacramento, CA: Legislative Analyst Office.

Osofsky, H. (2009). The intersection of scale, science, and law in Massachusetts v. EPA: State, national, and international approaches. In W. C. G. Burns and H. M. Osofsky, eds., *Adjudicating Climate Change State, National, and International Approaches*. Cambridge: Cambridge University Press, pp. 129–144.

Park, A. (2009). *Everybody's Movement: Environmental Justice and Climate Change*. Washington, DC: The Environmental Support Center, December. https://kresge.org/sites/default/files/Everybodys-movement-climate-social-justice.pdf

Rabe, B. (2009). Governing the climate from Sacramento. In S. Goldsmith and D. Kettl, eds., *Unlocking the Power of Networks: Keys to High-Performance Government*. Washington, DC: Brookings Institution, pp. 34–61.

Sayre, N. F. (2005). Ecological and geographical scale: Parallels and potential for integration. *Progress in Human Geography* 29(3), 276–90. https://doi.org/10.1191/0309132505ph546oa

Schlosberg, D., and Collins, L. (2014). From environmental to climate justice: Climate change and the discourse of environmental justice. *Wiley Interdisciplinary Reviews: Climate Change* 5(3), 359–374.

Stehr, N., and von Storch, H. (2005). Introduction to papers on mitigation and adaptation strategies for climate change: Protecting nature from society or protecting society from nature? *Environmental Science & Policy* 8(6), 537–40. https://doi.org/10.1016/j.envsci.2005.08.001

Stone, D. (2012). California tackles climate change, but will others follow? *National Geographic News*, November 16. www.nationalgeographic.com/history/article/121116-california-cap-and-trade

US Environmental Protection Agency. (2012). Environmental Justice Definition. Collections and Lists. US EPA. www.epa.gov/environmentaljustice

Williamson, R. (2015). Towards a multi-scalar methodology: The challenges of studying social transformation and international migration. In S. Castles, D. Ozkul, and M. Arias Cubas, eds., *Social Transformation and Migration*. Migration, Diasporas and Citizenship Series. London: Palgrave Macmillan, pp. 17–32. https://doi.org/10.1057/9781137474957_2

4

"When You Think about Climate Change, It Is a Social Justice Issue"

Between the Rock of Stability and the Hard Place of Politicization for Muslim Climate Actors

PAUL TOBIN, NAFHESA ALI, SHERILYN MACGREGOR, AND ZARINA AHMAD

The pursuit of significant greenhouse gas (GHG) emissions reductions is an urgent governance problem, necessitating transformative action. This volume contributes to understandings of a discord between two distinct and often antagonistic, yet not opposite, strategies and worldviews that underpin transformation efforts: those focused on stability and those on politicization. Here, we focus on two forms of stability and two of politicization. We suggest that in both schools of thought – those favoring stability and those for politicization – greater engagement is needed to understand and acknowledge existing patterns of structural inequality. In our chapter, we analyze the lived experiences of communities grappling with this discord, by learning from minoritized faith communities in the UK,[1] namely Muslim individuals pursuing greater climate action (see Hancock 2020; Koehrsen 2021; Tobin et al. 2023). Research to date on Muslim climate action has explored Muslim communities' climate initiatives within a variety of arenas yet has rarely found instances of strategic climate leadership from Muslim communities within "mainstream" fora (see DeHanas 2009; Gilliat-Ray and Bryant 2011; Koehrsen 2021; Nita 2014).[2]

In the UK, Canada, Germany, and many other Global North countries, Muslims are the largest religious group after Christians and atheists. Thus, the effective involvement of Muslim communities in the pursuit of countries' national climate goals, such as net zero plans, is paramount. Yet, despite this importance, scholarship must acknowledge the important role Muslim communities can play without

The support of the Economic and Social Research Council (ESRC) is gratefully acknowledged, having funded Paul Tobin via grant ES/S014500/1 during his involvement in this chapter. This chapter draws on work carried out as part of the "Towards Inclusive Environmental Sustainability" project, which is funded by a Leverhulme Trust Research Project Grant.

[1] We use "minoritized" rather than "minority" because it describes the active processes involved in differential allocations of power and resources between groups in society. The same active processes in the hands of more powerful white people lie behind the invisibilization and marginalization of people of color in the UK.

[2] We follow the understanding of Bell and Bevan (2021: 1208) that the "mainstream" environmental sector comprises more formal and resourced environmental organizations.

othering (i.e. treating as different or not belonging in some way) or instrumentalizing that participation (see Tobin et al. 2023). Muslims face multiple axes of inequality when participating in the climate policy process (Hancock 2015, 2020): Acknowledging *intersectional* inequalities (see Hill Collins 2019; Wilson 2013) is vital for understanding the realities of participation for Muslim communities. Islamophobia intersects with the impacts of racism (Meer and Modood 2019; Muslim Council of Britain 2016), economic inequality (Ali and Whitham 2021), and gendered discrimination (Zempi 2020) to affect Muslim communities' climate activities. In turn, these intersecting structures ensure that Muslim communities should not be considered a monolithic group. Hence, these intersecting inequalities have led to calls for *intersectional* climate justice that facilitates climate action while tackling the surrounding structures of injustice (Amorim-Maia et al. 2022; Mikulewicz et al. 2023).

What are the obstacles to the achievement of long-term, effective climate action via stability and politicization for Muslim communities in the UK? To inform our research, we conducted twenty-one semi-structured interviews between May 2021 and February 2022 with individuals based throughout the UK (listed in the chapter's Appendix).[3] Interviewees included nineteen Muslims who engaged in climate action from a wide variety of professions and roles and two non-Muslims who were involved in interfaith activities with Muslim communities. We deliberately interviewed a wide range of individuals to glean as broad an understanding as possible of the realities of Muslim experiences in climate action within the UK. Consistent with our intersectional approach to the analysis, we reflect on our own positionality as researchers who have designed the project and interpret the findings. We note the potential influence of the positionality of interviewers during the data collection process (see Fisher 2015; Merriam et al. 2001), which may shape our interpretations of the data, as well as the topics and scope of the conversations in the first place.[4]

Our analysis explores how the existing inequalities faced by Muslim communities mean that both stability and politicization alike pose challenges, meaning that it is only through a lens of climate justice that transformative climate action can be achieved. In doing so, we highlight that pursuit of long-term climate

[3] The UK hosted the United Nations Framework Convention on Climate Change (UNFCCC) Conference of the Parties (COP) 26 in October to November 2021, and there was increased awareness of climate change during this period.

[4] The first interviewer, A (Ali), conducted six interviews and identifies as a third-generation, British-born Muslim woman with Pakistani heritage. The second interviewer, B (Tobin), conducted fifteen interviews and is a British-born white man who is not a member of the Muslim community. We state which interviewer conducted each interview within the label for each interview; for example, "1B" was the first interview conducted by interviewer B. Finally, coauthors MacGregor and Ahmad did not conduct interviews; MacGregor is a white woman who is an immigrant to the UK and not a member of the Muslim community, while Ahmad is a British-born Muslim woman of South Asian heritage.

action for Muslim communities bears resemblance to being stuck between a rock and a hard place. Our findings are best encapsulated by a point made by one Muslim climate campaigner we interviewed: "we want political influence, but not necessarily through civil disobedience" (9B), yet such political protests have been one of the most high-profile means for climate campaigners to draw awareness to their cause.

In our short section on stability, we focus on two understandings of the concept as outlined in Chapter 1 of this volume. First, we analyze *stability as the status quo* by examining the impacts of structural inequalities and racism on Muslim communities' involvement in the UK climate movement (and beyond), and hence, the pursuit of *stability as lock-in* threatens to lock in these structural obstacles in the long term. This argument will be unsurprising considering the volume of work that already exists on inequalities within environmental action (Bell and Bevan 2021; Kaijser and Kronsell 2014; MacGregor 2009; Stephens 2020). However, politicization is not a straightforward solution. We analyze how the pursuit of politicization is, in itself, a complex and potentially dangerous process for Muslim communities due to the presence of, among other issues, harsher penalties in the justice system. By listening to the experiences of the Muslims interviewed for this project, we also aspire to represent a second form of politicization, namely *politicization as scholarly praxis*, as we wish for this chapter to represent some small form of intervention regarding existing inequalities. Hence, we draw explicitly from our interviewee data to center Muslim voices at the heart of our analysis. Reflecting on being between a rock and a hard place regarding stability and politicization outlined here, we argue that it is only through an overarching pursuit of intersectional climate justice that the benefits of the two schools of thought may be achieved. In short, as our title highlights, "when you think about climate change, it is a social justice issue" (10B).

Our chapter is structured as follows. First, we begin by outlining the existing research literature on Muslim communities as environmental actors, noting a dearth of studies that can identify Muslim climate actors being able to shape "mainstream" climate activities, despite environmentalism being a core concern of many Muslims. Second, we explore stability, by analyzing two manifestations of the strategy for Muslim communities in the UK. Third, we turn our focus to pursuing re/politicization. While we note that the goals of repoliticization are key to efforts to subvert existing power hierarchies, we emphasize the need for closer consideration of the realities of "othered" experiences to make these efforts effective. Fourth, we look beyond the stability/politicization discord to detail how considerations of intersectional climate justice must be at the heart of climate strategies. Fifth, we discuss "where next" for developing a more inclusive long-term response to climate change and conclude, highlighting three areas for future research.

4.1 Muslim Communities as Environmental Actors

Most people working in the "mainstream" environmental and climate sectors in the UK are white (Bell and Bevan 2021); there is little to zero data on the extent to which Muslim communities work in these sectors. In 2011, though, 92.2 percent of UK Muslims were People of Color (ONS 2011). As such, research on predominantly Muslim communities in the city of Manchester, UK, has shown that they face repeated injustice when liaising with white policymakers (MacGregor et al. 2022). Many Muslims see Islam as an explicitly *environmental* faith, in line with the sentiments favoring sustainability that feature within the *hadith* (discourses). Koehrsen (2021) summarizes the research to date on Muslims and climate action, noting that the UK and Indonesia predominate in the existing literature, but with different foci, as Islam is the majority religion in Indonesia but not in the UK. Minority faith groups' participation in climate policymaking – and indeed, wider policymaking – reflects different political dynamics to those of Christian communities, which are the largest faith group in much of the Global North and may even be formally affiliated to the institutions of the state, as is the case in the UK. In contrast, minority groups may have been "othered" within such societies (Modood 2019), which in turn affects their political participation.

Regarding Muslims in the UK, research has found that their minoritized status influences the extent to which communities can exert political influence, focusing on local rather than broader leadership or strategic roles. Gilliat-Ray and Bryant (2011) found in 2009–2010 that Muslims engaged in a variety of environmental conservation and sustainable horticulture initiatives, albeit in small numbers. Relatedly, DeHanas (2009) analyzed a radio show during Ramadan in 2007, exploring how guest presenters integrated their faith, environmental protection, and women's political representation into their hosting of the show. Nita's (2014) study of Christian and Muslim groups in the Transition Towns Movement during 2007–2010 found Muslims and Christians alike felt marginalized from environmental groups because of their faiths but also from their faith communities for engaging in explicitly green behaviors and activities. Finally, in a comprehensive volume examining the United States and the UK, Hancock (2018) analyzed Islamic environmentalism between May 2012 and July 2013, finding that at that time in the UK, campaigners engaged "in moderate forms of action that effect mostly individual change, not the radical systemic change necessary to realize their vision of the future" (Hancock 2018: 137).

There has been relatively sparse research on Muslim climate actors since those of the late 2000s and early 2010s, yet the years following have been marked by significant, high-profile activity on the part of the climate movement, necessitating further research into this area. Hence, research on Muslim communities in the UK has shown that they are environmentally active. Yet this research is increasingly out of date and does not explore the experiences of Muslim communities in seeking to play strategic roles in "mainstream" organizations, or how these experiences will shape the pathways employed to secure long-term climate action. Responding to these gaps, we examined the "between-the-scenes" roles of Muslims as "climate intermediaries" in the UK, finding that they were commonly invisibilized (i.e. treated as and made to feel invisible) and instrumentalized by "mainstream" environmental actors, and hence often collaborated through either interfaith work or within Muslim communities (Tobin et al. 2023). This chapter seeks to develop those arguments further, by examining the strategic dilemma in which UK Muslims find themselves because of widespread intersectional inequalities.

4.2 Stability

We focus on two forms of stability: stability of the status quo and, in turn, the dangers of "locking-in" policy solutions that would create a path-dependent route to entrenching these existing inequalities. As noted at the start of this chapter, the following section is shorter than that on politicization, as our findings align with existing research on structural inequalities within environmental action. Nevertheless, it is pertinent to explore these ideas further through the voices of members of Muslim communities, as different marginalized communities face unique structural barriers, and also in the context of considering a conceptual discord between the pursuits of stability and politicization.

4.2.1 Stability as the Status Quo

Muslims are the lowest-paid faith community in the UK (see Hancock 2018: 77–79). Multiple interviewees highlighted the high levels of poverty within Muslim communities in the UK as an obstacle to climate action (4B, 6B, 7B, 8B, 12B). As one interviewee (4B) lamented, "if you're a poor family, your first priority is not being environmentally conscious. Your first priority is eating and working and learning and getting your kids through school." This struggle ensures that for many Muslim families, participation in climate action feels an impossible task; this situation that has been exacerbated in the UK by more than a decade of government-led "austerity" measures during the 2010s and a cost-of-living crisis in the early 2020s.

Furthermore, and as introduced earlier, intersectional inequalities cut across financial hardships. "Being first/second generation immigrant communities – I think that really impacts how we act on it [climate change]. Secondly, again faith [Islamophobia], and thirdly, gender… those three intersect a lot in my voluntary capacity" (11B). Racism was also a frequent theme for interviewees (3B, 5B, 10B,

11B). For one interviewee wishing to support activities in the local park to become more climate-friendly (5B), the "direct abject racism" of the local council, including being told by a staff member "we don't want your lot running the park," meant that their ideas were blocked before getting off the ground. Hence, the pursuit of any form of policy stability that maintains this status quo for many Muslim communities would be unconscionable.

4.2.2 Stability as Policy Lock-In

Considering the existing context of intersecting inequalities faced by Muslim climate actors, locked-in trajectories hold the potential to entrench the above structural dynamics into long-term policymaking. Where policymakers do engage with minoritized communities, it is usually within depoliticized fora that have been designed to match the needs and assumptions of policymakers. Even if Muslim individuals have the capacity to join state-led initiatives, they may still be invisibilized while doing so, doubling the difficulty of climate action and damaging morale (5B, 10B, 12B; see Herzog 2018). Moreover, due to this lack of diversity, even individuals wishing to encourage more inclusive climate action can make suggestions that are not suitable for Muslim climate actors: "[to] the white, middle class, middle-aged individual who's usually engaged in environmentalism … it doesn't even enter their mind for there to be a barrier" (12B). Therefore, policies and activities are designed that cannot be accessed by Muslim communities, despite existing in principle to support marginalized communities.

Examples of policies designed without consideration of Muslim communities' needs include loan systems that are designed to pay for the installation of solar panels for mosques but are inaccessible due to their reliance on interest payments; grant application processes designed to encourage wider participation but that are too complex for communities to complete without expensive grant-writing support; and a lack of institutional support for gaining planning approval once funding has been won (1A, 4B, 6B; see also Tobin et al. 2023: 10). Collectively, such well-meaning yet ill-fitting policies reflect how the pursuit of "locked-in" stability, without wider representation within the policymaking process, will likely result in policies being locked in that cannot be used by their policy targets. In short, because of existing inequalities, and their potential to be locked in, none of our interviewees suggested that maintaining stability or the status quo would be beneficial paths to follow.

4.3 Re/politicization

Having briefly established the challenges posed by two forms of stability for Muslim climate actors, re/politicization offers a means of bringing critical perspectives into climate strategies. Re/politicization of climate action holds the potential for Muslim communities to shape the assumptions of policymakers behind the scenes, generating more effective policy outputs that may address intersecting inequalities. However, re/politicization efforts in the UK must also consider the lived realities of Muslims as othered communities, and the complex power inequalities therein. In this section, we begin by exploring the most high-profile politicizing strategies employed by the "mainstream" climate movement, namely protests and civil disobedience, finding that many Muslims are also reluctant to engage in activities that entail political conflict, due to their already marginalized positions. Yet, of course, the diversity within Muslim communities in the UK means that this sentiment is not universal. We then examine less politically charged means for Muslims to participate in climate change politics, finding that obstacles also exist in these arenas, too.

4.3.1 Protests, Marches, and Civil Disobedience

One of the most high-profile means for climate campaigners to politicize their cause has been through the usage of protests, demonstrations, and marches, such as the Fridays for Future school strikes movement led by Greta Thunberg (see Tosun and Debus, Chapter 2, this volume). More dramatically, others have pursued civil disobedience strategies, as seen through the disruption to sporting events led by, among others, Just Stop Oil (Kinyon et al. 2023). While there are more avenues for politicization available to campaigners than just protests, the high-profile and high-participation nature of protests means that this case is especially important for examination when discussing the realities of politicization. Many of our interviewees expressed disquiet at the thought of participating in these politicization strategies, for two main reasons.

First, multiple interviewees suggested that the spaces for protests and disruptions were created by people who appeared to be from a narrow range of privileged backgrounds, which limited the appeal of participation (3B, 4B, 7B, 12B, 14B). One Muslim climate organization leader (7B) wondered if "people like me didn't do this sort of stuff [protests and marches], because when I walked past, the many, many, many, many people who were there were all white and middle class." Here we see that the white-led nature of the environmental movement that we noted in Section 4.1 (see Bell and Bevan 2021) hinders involvement in "mainstream" activities for Muslims of color. Relatedly, the legacy of being from an immigrant family also influenced the willingness of our interviewees to participate in political protests: "I have very specific reasons for not taking part in protests. I was brought up in Pakistan … you would go there [to protests] and you would find that there were other ideas, or other demands and other things that had infiltrated that" said one interviewee (14B). As such, with protests being a dominant manifestation of climate action for many social movements, the design of protest activities by predominantly white campaigners in the UK, who bring their own understandings of what it means to protest, can make this form of activism exclusionary to Muslim communities. Indeed, one interviewee also noted the desire to avoid being involved in anything controversial, suggesting that being rebellious is counter to their desire to "just want to live a normal life, take care of their family" (4B), and indeed that many Muslim communities are conservative regarding disruption (also 6B, 9B).

Second, consolidating the difficulties of participating in protest activities is the influence of structural inequalities against Muslim communities (sustained by Islamophobia) that permeate the justice system. Interviewees discussed that institutional racism within the justice system added an extra level of inaccessibility to involvement in political protests (3B, 10B, 11B; see also Shankley and Williams 2020). Summarizing this sentiment, a Muslim business leader (3B) reflected:

> a lot of people were saying the XR movement was very much a white middle class movement. People that hoped to be arrested. And I know when I went down there, I wasn't keen to get arrested. Was I gonna wear a hooded top well?

This reflection on avoiding wearing a hooded top both to reduce the likelihood of being targeted by the police and to fit in with "mainstream" environmental actors featured in multiple interviews (3B, 4B, 10B, 12B, 14B). As a result, one Muslim climate campaigner (10B) who had previously participated in protests found that they could no longer take part in "actions" as they had become more disruptive: "We've been stopped, as young Asians, by police and treated differently compared to how we would have seen our white friends being treated." These reflections matter: Protests are one of the few means for environmental movements to influence policymakers, yet the messages these events carry are inherently streamlined due to the lack of diverse participation within them. "You will get off with a warning, I will get a criminal record" (11B) summarized one community activist.

Still, as explained in the introductory section of this chapter, let us not homogenize the Muslim experience in the UK. Other interviewees challenged the idea that Muslims would not feel welcome at protests. Indeed, the geographical origins of interviewees' familial heritage influenced this perspective. In contrast to the abovementioned interviewee whose family originated in Pakistan, another interviewee's family had come to the UK from Bangladesh: "we've got this huge, rich history of protest, [including] the role that women played … So this idea that we don't want to protest, it isn't a space for us, is nonsensical really" (11B). Furthermore, one Muslim climate campaigner (10B) who works at a "mainstream" environmental organization explained proudly the role they had played in co-planning the biggest climate march to date at that point, around the Paris climate conference in 2015.

Moreover, Extinction Rebellion has been supported by a specifically Muslim-led group, XR Muslims, since 2019, that has participated in protests. As a result, rather than querying whether protests can ever be accessible for Muslim climate actors, "[t]he question should be, 'what can we do to make our protests inclusive?' … [but] the question is never framed in that way. It's always about, 'well, why don't you come?'" (11B). As a result of this current inaccessibility, we must look also beyond protests and marches despite their status as one of the more high-profile forms of politicization.

4.3.2 Pursuing Politicization through Non-protest Channels

Many Muslim climate actors pursue other strategies for politicizing climate policymaking beyond protests (3B, 7B, 9B, 10B, 12B). For instance, one interviewee seeks to combine

> the deeply pragmatic and the deeply radical. Radical in the change that I want to see … But pragmatic in understanding that … we need people pulling in every single direction, because actually there's no one route to social change … [as well as] join marches, join activist circles, do civil disobedience … we need everybody and we need lots of theories of change. (10B)

As a result, we found a variety of approaches in play. Muslim climate actors have tried to build coalitions with other communities to then influence policymaking through more institutionalized means, particularly by working with other minoritized groups, such as LGBTQ+ activists (7B, 12B), Indigenous communities (7B), and other faith groups, such as Sikhs (3B, 7B, 9B, 12B). Indeed, this interfaith work enables Muslim communities to collaborate with like-minded actors while potentially benefiting from greater access to resources (12B; see also Tobin et al. 2023). In line with these values, for example, the environmental nongovernmental organization (ENGO) Muslims Declare was established in 2021 with the goal of prioritizing participatory events that bring together business leaders, politicians, experts, faith leaders, campaigners, and individuals through public fora, exhibitions, assemblies, and workshops (Muslims Declare 2023).

Nevertheless, despite the potential benefits afforded by coalition building, interviewees also expressed a preference for working in Muslim-only spaces, because otherwise, they "can't find a safe space" (3B). One Muslim climate campaigner (9B) explicitly sought to avoid including any form of civil disobedience in the activities of their Muslim-only organization, and instead they try to build from a common set of shared understandings: "it's trying to find a narrative that really connects with the worldview that most Muslims are in within the UK. Their precarity, their sense of fear" (9B). Muslims' unique positionality within UK climate politics means that because “we can’t engage with mainstream organizations, we’re just going to do our own” (12B). Building on this point, another Muslim climate campaigner noted that “I would rather play my part with my own small group that I can influence, and hopefully it will eventually have that ripple effect and impact more people. But I’m not prepared to go out and do protests and things” (14B). Yet this approach resonates with the realities explored earlier regarding stability as the status quo; by feeling forced to undertake climate politicization in a siloed community, Muslim climate actors possess limited resources and face limited access to influential policymakers, which in turn perpetuates the lack of representative or effective policies that can then be locked in without Muslim involvement. Hence, as we explore next, many of our interviewees emphasized that their activities need to be understood as part of a wider campaign pursuing intersectional climate justice, regardless of the exact strategy in place.

4.4 Next Steps: The Pursuit of Intersectional Climate Justice

Many of our interviewees brought up the concept of justice without it having been suggested by the interviewer (2B, 3B, 7B, 9B, 10B, 11B, 12B, 13B). A Muslim leader of a not-for-profit climate organization noted that climate change is only one part of a broader conversation: “I see my work on climate as sitting within this larger, package of progressive social justice causes” (7B); for another, “I don’t think about climate change as an environmental issue anymore” (10B). Interviewees discussed the need to bring in climate justice together with wider issues such as inequality toward LGBTQ+ communities (7B), gender inequality (7B, 9B, 10B, 11B), and racism (2B, 3B, 5B, 10B, 11B). These areas of injustice were understood to be differing manifestations of the same coloniality and oppression (3B, 5B, 7B, 9B). Indeed, this oppression is arguably the root cause of the obstacles to participation in climate action.

> When you look at climate change protests, I would say it isn’t our shared cultural heritage that is really a barrier … I think it is racism. And I don’t think there’s enough anti-racism practice or approach into how environmental sustainability, resistance, activism, or protesting is practiced. (11B)

That is, by encompassing climate action *within* wider justice issues and tackling those at the same time, subsequent changes can then be made that facilitate, for example, protests, which in turn makes more high-profile disobedience tactics for politicization available to Muslim communities.

Of course, as with each of the areas discussed, there is diversity of opinion within Muslim communities regarding the role for justice and how to manifest its importance. “Even within the Muslim climate sphere there might be those of us [where] the model is conservation and nature, and for some of us the model is justice" (7B). Indeed, this distinction in priorities brings its own disagreements and discord. One campaigner noted that for their organization, it became very difficult to strategize for environmental campaigns because some wished to stick to the science-based nature of climate change while others wished to focus on oppression and justice (9B). Lastly, there was also frustration that when white and non-Muslim allies in the climate movement did shift their attention to climate change as a justice issue, there was obliviousness regarding their complicity in these structural dynamics (11B). As such, even efforts toward championing greater engagement with justice issues as a means of bridging any divide between stability and politicization bring their own delicate areas for reflection, engagement, and action.

4.5 Conclusion: Where Next for Muslim Climate Actors and for Research?

Our chapter has shown how the pursuit of stability may perpetuate existing inequalities and encourage the locking-in of climate strategies that Muslim communities are excluded from shaping, resulting in effectual policy design. However, in pursuing repoliticization strategies, many interviewees highlighted experiencing similar such inequalities. Engaging in political activities as a Muslim citizen/individual in the UK is rife with difficulty. In the case of political protest, institutionalized discrimination ensures that any civil disobedience activities by Muslims result in a higher probability of arrest, and worse conditions following such an arrest. Due to these lived experiences, the appeal of political conflict must be weighed against a more vulnerable strategic position. Our findings support the argument that wider consideration of justice issues is needed within climate policy design; as leadership (slowly) diversifies from the current status quo, policy structures that support increasing critical engagement from invisibilized communities will enable more representative and effective policies to be designed. Without this expectation, policy design emphasizing "policy stability" may lock in the predominant assumptions around engagement and representation, which currently marginalize large numbers of people. We thus bring the realities of depoliticized climate politics as faced by invisibilized communities into conversation with one another and analyze the impacts of minoritization in shaping Muslim communities' strategies for climate action.

Despite these obstacles to participation, our research with Muslim communities also highlighted myriad instances of climate action. These activities included national-level strategic oversight that had rarely been identified by the existing literature that was mostly carried out in the late 2000s and early 2010s. We suggest three research questions for future investigations that may build on our work. First, to what extent are activism and advocacy around climate change in Muslim-majority contexts relevant or motivating for Muslims in the UK? Second, which initiatives have been especially successful in encouraging Muslim individuals to design policy, and how can such strategies be adapted and applied in other contexts? Finally, to what extent have these initiatives involved deliberate and official anti-racism or decolonizing strategies on the part of white-dominated institutions?

Appendix: Descriptions of Interviewees

1A: June 22, 2021, Muslim councillor.
2A: June 18, 2021, Muslim resident and local campaigner (climate volunteer).
3A: July 13, 2021, Muslim climate organization movement builder.
4A: June 18, 2021, Muslim mosque trustee, solicitor, and chair for local council of mosques.
5A: October 12, 2021, Muslim councillor and executive member.
6A: July 5, 2021, Muslim councillor.
1B: May 18–19, 2021, city-level not-for-profit employee.
2B: October 21, 2021, Christian faith leader and city-level interfaith board member.
3B: November 1, 2021, Muslim business leader.
4B: November 2, 2021, Muslim faith representative leader.
5B: November 10, 2021, Muslim local organizer.
6B: November 18, 2021, Muslim local organizer.
7B: December 1, 2021, Muslim not-for-profit climate organization leader.
8B: December 1–13, 2021, Muslim business leader in the environmental sector.
9B: December 13, 2021, Muslim climate campaigner.
10B: January 14, 2022, Muslim climate campaigner and organization leader.
11B: January 14, 2022, Muslim community activist and enabler.
12B: January 19, 2022, Muslim community leader.
13B: January 19, 2022, Muslim faith representative and climate spokesperson.
14B: January 31, 2022, Muslim climate volunteer.
15B: February 2, 2022, Muslim civil servant.

References

Ali, N. and Whitham, B. (2021). Racial capitalism, islamophobia, and austerity. *International Political Sociology* 15(2), 190–211.

Allan, J. I. (2020). *The New Climate Activism: NGO Authority and Participation in Climate Change Governance*. Toronto: University of Toronto Press.

Amorim-Maia, A. T., Anguelovski, I., Chu, E., and Connolly, J. (2022). Intersectional climate justice: A conceptual pathway for bridging adaptation planning, transformative action, and social equity. *Urban Climate* 41, 101053.

Bell, K. and Bevan, G. (2021). Beyond inclusion? Perceptions of the extent to which Extinction Rebellion speaks to, and for, Black, Asian and Minority Ethnic (BAME) and working-class communities. *Local Environment* 26(10), 1205–1220.

De Cordier, B. (2009). Faith-based aid, globalization and the humanitarian frontline: An analysis of Western-based Muslim aid organizations. *Disasters* 33(4), 608–628.

DeHanas, D. N. (2009). Broadcasting green: Grassroots environmentalism on Muslim women's radio. *The Sociological Review* 57(2), 141–155.

Fisher, K. T. (2015). Positionality, subjectivity, and race in transnational and transcultural geographical research. *Gender, Place and Culture* 22(4), 456–473.

Gilliat-Ray, S. and Bryant, M. (2011). Are British Muslims "Green"? An overview of environmental activism among Muslims in Britain. *Journal for the Study of Religion, Nature and Culture* 5(3), 284–306.

Haluza-DeLay, R. (2014). Religion and climate change: Varieties in viewpoints and practices. *Wiley Interdisciplinary Reviews: Climate Change* 5(2), 261–279.

Hancock, R. (2015). "Islamic" environmentalism in Great Britain. In T. Peace, ed., *Muslims and Political Participation in Britain*. Abingdon: Routledge, pp. 103–123.

Hancock, R. (2018). *Islamic Environmentalism: Activism in the United States and Great Britain*. Abingdon: Routledge.

Hancock, R. (2020). Muslim environmentalists, activism, and religious duty. In M. Peucker and M. R. Kayikci, eds., *Muslim Volunteering in the West*. Cham: Palgrave Macmillan, pp. 141–160.

Herzog, B. (2018). Invisibilization and silencing as an ethical and sociological challenge. *Social Epistemology* 32(1), 13–23.

Hill Collins, P. (2019). The difference that power makes: Intersectionality and participatory democracy. In O. Hankivsky and J. S. Jordan-Zachery, eds., *The Palgrave Handbook of Intersectionality in Public Policy*. Cham: Palgrave Macmillan, pp. 167–192.

Kaijser, A. and Kronsell, A. (2014). Climate change through the lens of intersectionality. *Environmental Politics* 23(3), 417–433.

Kinyon, L., Dolšak, N. and Prakash, A. (2023). When, where, and which climate activists have vandalized museums. *npj Climate Action* 2(1), 27.

Koehrsen, J. (2021). Muslims and climate change: How Islam, Muslim organizations, and religious leaders influence climate change perceptions and mitigation activities. *Wiley Interdisciplinary Reviews: Climate Change* 12(3), 702.

MacGregor, S. (2009). A stranger silence still: The need for feminist social research on climate change. *The Sociological Review* 57(2 suppl), 124–140.

MacGregor, S., Ali, N., Ahmad, Z. et al. (2022) Better policy through symmetry? Examining the whiteness of green governance in Manchester. Proceedings of the Nordic Environmental Social Science (NESS) Conference, Gothenburg, June 7–9. https://research.manchester.ac.uk/en/publications/better-policy-through-symmetry-examining-the-whiteness-of-green-g

Meer, N. and Modood, T. (2019). Islamophobia as the racialization of Muslims. In I. Zempi and I. Awan, eds., *The Routledge International Handbook of Islamophobia*. Abingdon: Routledge, pp. 18–31.

Merriam, S. B., Johnson-Bailey, J., Lee, M. Y. et al. (2001). Power and positionality: Negotiating insider/outsider status within and across cultures. *International Journal of Lifelong Education* 20(5), 405–416.

Mikulewicz, M., Caretta, M. A., Sultana, F. and Crawford, N. J. W. (2023). Intersectionality and climate justice: A call for synergy in climate change scholarship. *Environmental Politics* 32(7), 1275–1286.

Modood, T. (2019). *Essays on Secularism and Multiculturalism*. London: ECPR Press.

Muslim Council of Britain (2016). *British Muslims in Numbers: A Demographic, Socio-economic and Health Profile of Muslims in Britain Drawing on the 2011 Census*. London: Muslim Council of Britain. www.mcb.org.uk/wp-content/uploads/2015/02/MCBCensusReport_2015.pdf

Muslims Declare (2023). Our story. Website. www.muslimsdeclare.org/about/

Newell, P. (2006). *Climate for Change: Non-state Actors and the Global Politics of the Greenhouse*. Cambridge: Cambridge University Press.

Nita, M. (2014). Christian and Muslim climate activists fasting and praying for the planet: Emotional translation of "dark green" activism and green-faith identities. In R. G. Veldman, A. Szasz and R. Haluza-DeLay, eds., *How the World's Religions Are Responding to Climate Change: Social Scientific Investigations* (1st ed.). Abingdon: Routledge, pp. 229–243.

ONS (Office for National Statistics) (2011). Muslim population in the UK. www.ons.gov.uk/aboutus/transparencyandgovernance/freedomofinformationfoi/muslimpopulationintheuk/

Shankley, W. and Williams, P. (2020). Minority ethnic groups, policing and the criminal justice system in Britain. In B. Byrne, C. Alexander, O. Khan, J. Nazroo and W. Shankley, eds., *Ethnicity, Race and Inequality in the UK: State of the Nation*. Bristol: Bristol University Press, pp. 51–72.

Stephens, J. C. (2020). *Diversifying Power: Why We Need Antiracist, Feminist Leadership on Climate and Energy*. Washington, DC: Island Press.

Tobin, P., Ali, N., MacGregor, S. and Ahmad, Z. (2023). "The religions are engaging: Tick, well done": The invisibilization and instrumentalization of Muslim climate intermediaries. *Policy Studies* 44(5), 627–645.

Wilson, A. R. (2013). *Situating Intersectionality: Politics, Policy, and Power*. New York: Palgrave Macmillan.

Zempi, I. (2020). Veiled Muslim women's responses to experiences of gendered Islamophobia in the UK. *International Review of Victimology* 26(1), 96–111.

Part II

Political Economy

5

Politicizing Coal Burning

Phaseout Policies from Cheap Signals to Emergent Norms and North–South Contention

STACY D. VANDEVEER

After generations as one of the primary sources of electricity generation across much of Europe and North America, since about 2013 coal burning has more recently emerged as among the most contentious areas of energy and climate change policy in the much-discussed, but little-practiced, national- and global-scale energy transitions. Energy policymakers tend to prefer stable, longer-term policies and planning to regulate electricity generation (and some aspects of pricing). However, accelerating climate change impacts, the increasing urgency of climate action, incremental increases in many public sector greenhouse gas (GHG) emission reduction goals, several changes in fossil gas and electricity generation prices and markets, and a growing inclination of civil society actors to target coal burning explicitly have combined to push an increasing number of jurisdictions, firms, and multilateral organizations to phase out supports for existing or future coal generation, and/or to announce a specific deadline by which coal burning is to be eliminated and banned.

Efforts to explicitly politicize coal burning, and thereby disrupt energy policymaking and energy investment markets, to achieve phaseouts of coal burning are, in some spaces, leading to a set of increasingly stabilized policy norms associated with phasing out and permanently ending or banning coal burning. As coal phaseout dates become common – or "normed" – in some jurisdictions and organizations, contention about coal and other fossil fuel policies shifts debates, discourses, and activism in local and transnational energy and climate politics. In short, the establishment and proliferation of coal phaseout discourses and policy norms also shape growing backlash politics and contention in reaction to coal phaseout demands in domestic and transnational politics.

This chapter brings together three intersecting discussions and developments. First are the set of questions at the center of this volume around policy stability and politicization posed by Paterson, Tobin, and VanDeveer (2022). Second is the relatively sudden appearance and transnational diffusion/proliferation of coal phaseout policy demands since about 2015, including announcements and (sometimes) enactments by various public and private sector organizations at multiple levels of authoritative scale from the global to local (Adekoya et al. 2023; Jakob and Steckel 2022; Misik and Pracharova 2023; Ohlendorf et al. 2022; Rentier et al. 2019; VanDeveer and Boersma 2022; Vinichenko et al. 2023). Third is the long-standing social science literature on norms and normative change in comparative, transnational, and global politics (see, e.g., Finnemore 1996; March and Olsen 1989; Mitchell and Carpenter 2019; Nadelman 1990; O'Neill et al. 2004; VanDeveer 1997). The guiding question posed here is as follows: What can we learn about the policy stability and politicization dynamics, and about comparative and transnational normative change, through an exploration of transnational coal phaseout politics between 2013 and 2024? This analysis largely sets aside earlier national coal phasedowns and phaseouts in countries such as the UK and the Netherlands, which took place in different political and social contexts and were driven by a number of different causal factors. Put bluntly, it was not coal burning and the resulting contributions to global climate change at which Margaret Thatcher's considerable discursive and policymaking powers were aimed.

As the title suggests, as declared coal phaseout policy norms gained popularity in the 2010s, often pushed by climate activist organizations, the early adopters were often states which had already mostly phased out the coal sector for non-climate change reasons, or which had never actually used much coal in electricity production. These declarations might be seen as "cheap signals" of climate change leadership or climate virtue. But as coal phaseout enters domestic and transnational politics and begins to impact more coal-dependent societies and economies, politics becomes considerably more contentious and normatively complex and fraught.

The chapter's six sections proceed as follows: First is a brief summary of some major aspects of coal's history as both a major source of energy and a major source of fractious national and transnational politics that is at the heart of the chapter. In the second and third sections, I turn to discussions of coal politics at United Nations Framework Convention on Climate Change (UNFCCC) Conferences of the Parties (COPs) and then to a short explanation of the usual content of coal phaseout policy norms. Here, we see the push by climate change mitigation activists, experts, and organizational leaders to explicitly politicize the burning of coal and demand date-certain deadlines for the cessation of coal burning and the financing of coal mining and coal burning facilities. The fourth and fifth sections explore the emergence of a set of coal phaseout norms in domestic and transnational politics and the increasing contestation or repoliticization of these emerging norms, respectively. The chapter concludes by drawing out four sets of implications for coal and fossil fuel phaseout advocates and skeptics, for understanding antagonisms between stability and politicization and for comparative environmental, energy, and climate change politics.

5.1 From Black Diamonds and the Big Smoke to Planet Killer

Humans have been burning coal for thousands of years, across continents and within ancient empires associated with China, Greece, Rome, and Aztec histories – to name only a few. But it is Industrial Age expansion – with the advent and rapid proliferation of coal-fired steam engines and coal-fired power plants (first erected in London in the 1880s) – that cements coal as a major industrial energy source from the late nineteenth century, throughout the twentieth century, and into the early twenty-first century. Throughout much of this period, it was labor rights, labor movements, and worker health and safety that often dominated national politics in states with large coal reserves and significant coal mining and coal burning. Mitchell's (2013) work argues that coal's fueling of industrialization and huge labor unions is a centrally important driver of democratization in industrial societies. Another illustration of coal's central importance can be found in the fact that today's European Union has its origins in the 1951 European Coal and Steel Community, whose early priorities included more housing for coal miners (Merry 1955).

In the latter half of the twentieth century, environmental movements and environmental policies – especially those focused on air pollution and its risks to human health, ecological/nature protection, and architectural and aesthetic damage – began to reframe coal burning as a source of dangerous and expensive pollution-induced harms. Just as it is difficult to explore varieties of capitalism, comparative labor rights, and labor movements – or the comparative histories of socialism and socialist states – without reference to coal mining and coal burning, so too is it unlikely that comparative histories of European and North American environmental movements and policymaking could ignore coal.

Since 2015, as climate change accelerates and consensus about the needs to act to mitigate climate change stabilized, we see another dramatic shift toward framing coal burning as "bad" and "unjust" from the global to local scale (Boersma and VanDeveer 2016a, 2016b; Jakob et al. 2020; Selin and VanDeveer 2025; VanDeveer and Boersma 2022). Coal is now often framed by environmental advocates as the worst or most dangerous fossil fuel, because it is the most carbon-intensive fossil fuel and also the fossil fuel that often produces the most visible, local damage to human and ecological health.

From the early 2000s, pro-coal rhetoric about "clean coal" and "wars on coal" emerged alongside a politics of "coal phaseout" demands and announced policies. Civil society actors have demanded that public and private actors phase out coal burning and ban it – and divest entirely from coal-related investments. Two prominent examples of such activism are the Climate Action Network's several years of anti-coal burning advocacy, including as part of their "Fossil of the Day" awards at UNFCCC COPs,[1] and the many environmental organizations and anti-coal burning campaigns supported by Bloomberg Philanthropies, including the Bloomberg Global Coal Countdown.[2] Across most Organisation for Economic Co-operation and Development (OECD) states, coal is increasingly framed in political discourse as the worst offender in climate change politics.

Many public sector actors at national, provincial, and municipal scales have either declared themselves "coal-free" or publicly announced deadlines/dates by which coal will no longer be burned in their jurisdictions. Many private and multilateral financial institutions have announced they will no longer fund or invest in coal burning facilities. These shifts have several empirical and discursive origins, including the failure (through at least 2024) to reverse global growth in GHG emissions and the frustration and anger about this fact among most climate change activists and scientists, in combination with ethics and justice-related aspects of contemporary climate change discourses and the increased price competition from renewable energy (VanDeveer and Boersma 2022).

5.2 Coal and the COPs

One place to identify and explore shifting discursive and policy norms related to coal burning is in public debates, state and non-state actors' position statements, and negotiated UNFCCC documents around the annual UNFCCC COPs. While environmental and climate activists and scientific analyses long focused attention on coal's contributions to climate change and the need to reduce GHG emissions from coal in some countries' domestic politics since the 1990s, the transnational normative shifts related to coal burning became much more apparent in the run-up to the 2015 UNFCCC climate change accord negotiations at COP21 in Paris. While the early 2000s saw "clean coal" exhibits and other explicit attempts to frame coal as an energy source that could be made climate-friendly, coal emerged as a priority target of decarbonization activists in the 2010s. The 2015 Paris Agreement – hailed by many as a political success in global climate change politics (Dimitrov 2016) because of a host of institutional advances achieved in the negotiations – makes no mention of the need to curb, much less phase out, coal or any other fossil fuels. Subsequent analysis about the more ambitious mitigation policies required to achieve the Paris Accord's goal of limiting global warming to 2.0°C or 1.5°C

[1] See Climate Action Networks' "Fossil of the Day" webpage: https://climatenetwork.org/resource_type/fossil-of-the-day/.
[2] For the Bloomberg Global Coal Countdown, see https://bloombergcoalcountdown.com/.

regularly demonstrated the need for a fairly rapid phaseout of coal burning, along with steep curbs in consumption of other fossil fuels (see, e.g., Friedlingstein et al. 2022; Vinichenko et al. 2023).

At COP23, in Bonn in 2017, the UK and Canadian governments launched a multistakeholder initiative called the Powering Past Coal Alliance (PPCA) that initially included twenty-seven state, subnational, and non-state actors endorsing phasing out (thermal) coal burning for electricity production globally by 2050 (Blondeel et al. 2020). Individual members of the Alliance can set their own phaseout date, before 2050. In only a month, membership had grown to fifty public and private sector organizations. PPCA has continued to grow in membership and ambition through COP28 in Dubai, where the United States announced that it would join the Alliance. Importantly, many early joiners had little or no coal in their energy portfolios when they announced the expected dates by which they would be coal-free. Examples include the UK, Sweden, California, and Oregon, for example. One might characterize the messages being sent by most of these early joiners as "cheap signals" (as the chapter title suggests). But evidence suggests that more authoritative policy norms are emerging as more coal-dependent actors – especially states – announce their membership. Such examples include Indonesia, Poland, and the United States, all of which had endorsed "Powering Past Coal" by 2023. As often happens with normative change over time, more actors begin to take more concerted policy action when the logic of appropriateness and the logic of consequences essentially point in the same direction (VanDeveer and Boersma 2022; Blondeel et al. 2020; Green 2018a). As more coal-dependent actors sign on, the "politics" is likely to get harder even as the anti-coal burning policy norms stabilize. As the norm becomes more influential, established, and authoritative, opposition by impacted local, national, and transnational actors also seems likely to intensify among the "holdouts" or opponents of planned coal phaseouts.

At COP26, in Glasgow in 2021, numerous state and non-state actors pushed for the formal declaration to include mention of the need to phase out coal. In the end, the declaration noted the need to "phase down" coal consumption, but no deadlines or rates of decline were included. Van Asselt and Green (2023: 2) argue "that the developments at and surrounding COP26 show that [anti-fossil fuel norms] are increasingly being adopted and institutionalized." The very public disagreements among states and among civil society actors at COP28 in Dubai in 2023, around what language to use in relation to fossil fuels, illustrate that as such norms gain influence globally, they provoke more opposition across scales from global to local.

As domestic and transnational coal politics change, environmental justice and injustice discourses and activism have added a moral weight – and increasingly morally charged and contested discourses – to both domestic and transnational coal phaseout politics over the last decade. It must be clear, however, that virtually all of these developments in policymaking debates and discourses remain highly contested once the norm and its related discourses move beyond the sending of cheap signals. COP28 illustrated that demands for phaseout goals and deadlines for coal and other fossil fuels continue to shape politics at the COPs, but that opposition to such a phaseout goal among most fossil fuel-producing states continues to keep phaseout language out of official multilateral declarations at the conclusion of each COP. UN Climate Change Executive Secretary Simon Stiell noted, in his closing speech at COP28, that "[w]hilst we didn't turn the page on the fossil fuel era in Dubai, this outcome is the beginning of the end."[3] This was his assessment of states' negotiators leaving out calls to "phase out" coal and other fossil fuels but including references to phase down coal and transition away from fossil fuels in a just manner.

5.3 Politicizing Coal Burning: Demanding and (Sometimes) Achieving Coal Phaseout Policies

What newly emergent policy norms can we identify across actor types, sectors, and scales? Coal has nearly always been political in some way, but who was politicizing it, and why, has varied substantially across time and space. Over 150 years, many countries' industrialization, political development, and labor histories are inseparable from coal burning and coal mining (Freese 2003; Malm 2016a; Mitchell 2013; Paxman 2022). By the last decades of the twentieth century, coal's substantial contributions to air pollution – and the negative ecological, aesthetic, and labor and public health impacts of such pollution – often placed the emissions from coal burning in the pantheon of domestic and transnational environmental discourses, politics, and environmental policymaking. For several decades, such politics focused on the emissions from coal burning, with environmental policymaking focusing on mitigating particulate and toxic emissions, or "scrubbing" sulfur dioxide emissions, and so on. Burning coal to generate electricity (and/or heat) was, for a while, assumed to continue.[4]

But more recent climate change and environmental justice activism politicizes coal *burning* itself, in conjunction with the resulting GHG, toxic, and particulate emissions and local and global inequities and injustices associated with impacts on people and communities. This politicization was/is explicitly linked to frustration, fear, and anger about the slow pace of GHG reductions and the inadequate outcomes of decades of policymaking (Anderson et al. 2020; Karlsson and Gelik 2020; Stoddard et al. 2021), the influence-peddling/corruption associated with coal

[3] See the UNFCCC website: https://unfccc.int/news/cop28-agreement-signals-beginning-of-the-end-of-the-fossil-fuel-era

[4] While coal burning is used for several industrial purposes, and for home/local heating, most coal phase-out politics is primarily concerned with burning coal for electricity production.

and other fossil sectors, and the notion that "we are running out of time" (de Moor 2023) – to name only a few themes embedded in anti-coal activism.

Recent work on the comparative politics of coal and energy policy, and the transnational and international politics of coal and energy policy, demonstrates dynamic combinations of transnational discourses and policy ideas and norms in complex interaction with highly diverse political and policymaking outcomes (VanDeveer and Boersma 2022; Jakob and Steckel 2022). Only a few years ago, it was possible to suggest that scholarship on global climate change politics and governance had paid little attention to emerging "global moral norms" (Green 2018a). Since then, however, numerous academic treatments of coal phaseouts movements, anti-fossil fuel activism, and multilateral agreements began to regularly characterize these policy ideas as emergent anti-coal or anti-fossil fuels norms and to discuss their proliferation (Blondeel et al. 2019; Green 2018b; van Asselt and Green 2023). Generally speaking, this literature agrees that norms are standards of behavior expected of certain actors – standards associated with a moral justification and/or the logic of appropriateness of particularly behaviors (Finnemore and Sikkink 1998; March and Olson 1989). Such scholarship also frequently borrows the concepts of "norm entrepreneurs" or "norm champions" – concepts denoting organizational and individual leaders or agents who knowingly seek to champion normative change – from norms scholarship, as well.

The individuals and organizations who founded and helped to build (and fund) the PPCA illustrate the importance of such entrepreneurial champions. In reference to public, private, and civil society actors' inclination to set and announce particular courses of action, or "policies" governing their own and possibly others' behaviors over time, the term "policy norm" is often used. Such literature and my ongoing research examine the discourses and demands of anti-coal and anti-fossil fuel activists, organizations, and social movements – and the emergent and growing set of enacted policies by public and private organizations – to specify a set of emergent policy norms.

I identify four major coal phaseout norms that appear most common among activist demands in transnational and comparative anti-coal politics, and the resulting announced phaseout policy initiatives by public sector actors and private financial institutions:

1. Ban/end the construction of new coal burning facilities.
2. Ban/end the financing of new coal burning facilities.
3. Publicly declare the goal of phasing out coal burning and stipulate a deadline by which it must occur.
4. Enact some type of official policy and periodic public reporting process to allow tracking of progress toward the phaseout goal.

Importantly, these policies – or policy norms – can be pursued by public, private, or civil society organizations and various levels of scale. So, for example, national or subnational public sector jurisdictions may pursue some or all of these – as can financial services firms, energy corporations, or energy-producing nonprofits. Importantly, bans on coal mining and coal exports – as with bans of oil and gas exploration and extraction – are often debated or demanded by some civil society actors but seem to be less commonly enacted to date. Lastly, note that the fourth policy norm on the list above is less a statement of a new policy norm and more of an implementation norm that helps make the policy declarations more credible and trackable over time.

5.4 Norm Making and Proliferation: Politicizing Coal Burning and Stabilizing Bans and Phaseout Deadlines

While previous sections summarize the changing coal politics and discourses in UNFCCC COPs since 2015, coal phaseout norm stabilization and proliferation are perhaps most apparent in public policies announced at the national level, among a growing set of subnational public actors, and among multilateral and private financial institutions. While normative changes associated with coal burning are far from universal, they are happening in many places and spaces, across local to global scales.

By early 2024, Beyond Fossil Fuels lists twenty-three European states with declared coal phaseout policies,[5] including three that have become "coal-free" since the 2015 Paris Agreement, thirteen with coal phaseout deadlines of 2030 or earlier, and seven with announced deadlines after 2030. Their data include another ten countries that historically had little or no coal in their electricity generation mix. In Europe, coal phaseout and coal-free declarations began with states using little or no coal but have moved over time to include more states in Central, Eastern, and Southeast Europe where significant coal burning continues. The coal phaseout norm is spreading. Furthermore, while the 2022 Russian invasion of Ukraine and the accompanying energy price spikes raised concerns about Europeans' commitments to climate change policies and renewable energy promotion, 2023 saw record levels of renewable energy and dramatic declines in European coal consumption. Ember (2024) found that coal power generation fell by more than 25 percent and European GHG emissions from the power sector fell by 19 percent. The report noted that "Europe's coal phaseout gathers pace" as about 20 percent of Europe's coal plants are set to close in 2024 and 2025, especially in Germany, Italy, Poland, Greece, and Spain. Coal power generation was cut by about half between 2016 and

[5] See Beyond Fossil Fuels' "Europe's Coal Exit" webpage: https://beyondfossilfuels.org/europes-coal-exit/.

2023. Given the increasingly ambitious GHG reductions goals expected by EU institutions and most member states, coal's decline in much of Europe seems likely to continue. Given that the Russian invasion of Ukraine generated urgent needs to move away from Russian natural gas among several countries with ready access to coal, the continued phasedown and phaseout policies and trends remain notable.

US coal consumption has been declining rapidly since the early 2000s, with at least ten states becoming "coal-free" by 2023. To be clear, a host of causal factors are driving coal consumption declines in the United States and Europe (VanDeveer and Boersma 2022), including market and pricing dynamics, as well as environmental and climate change movements, policies, and litigation. US and European coal politics illustrate complex combinations of moral responsibility, domestic political dynamics, and the importance of the changing economic prices related to coal burning and lower-carbon energy sources (VanDeveer and Boersma 2022). But in many respects, it has been easier (and common) for North American and European governments to talk the talk of coal phasedowns or phaseouts – and of GHG reductions and global responsibilities, more generally – than it has been for them to walk the walk of more significant domestic policy action and implementation.

Nevertheless shifts away from coal burning and toward more renewable energy (and some additional natural gas) continue. By 2024, the PPCA included almost sixty states, with about half of those located outside of Europe. The Alliance lists dozens more subnational governments and large corporations who also joined. Outside of Europe, the steepest declines in coal burning by late 2023 were in the United States, Chile, and Israel (Jaeger 2023). All three joined the PPCA as domestic declines in coal burning accelerated. At least five Latin American states where coal is currently burned to generate electricity also joined the Alliance in recent years. This suggests that coal phaseout norms are becoming increasingly common across the Western hemisphere.

But the growing influence of an emerging set of policy norms should not be mistaken for current or future universal consensus. In fact, the growing impact of coal phaseout norms seems likely to continue to engender substantially more coal-related contention. Despite seemingly constant climate policy debates, the International Energy Agency estimates that global GHG emissions hit a record high in 2023 (Twidale 2024) and the US Energy Information Agency estimates that China and India will account for about two-thirds of global coal consumption between 2022 and 2050 (US EIA 2024). As more states phase out coal burning, or foreclose the option of building a coal sector, the "holdouts" seem likely to receive more attention and political and economic pressure. Such holdouts might be large coal consumption states or large coal mining and coal exporting states (such as Australia). Also, after current coal burning states announce a coal phaseout deadline in international venues, as Vietnam and Bulgaria have done, political contention often shifts to domestic and international actors involved in implementing such goals (Do and Burke 2023; Spasic 2024). The road ahead for coal phase-out declarations and implementation seems unlikely to be linear, depoliticized, or lacking in contention (Medzhidova 2022; Muttitt et al. 2023).

The Obama and Biden US presidential administrations, together with influential EU and EU member state officials, used their traditionally substantial leverage over World Bank and International Monetary Fund (IMF) policies to push these organizations and other multilateral development banks to curtail coal financing. Since 2013, the World Bank officially has a moratorium on directly financing coal burning power plants, and more recently it supports "fair energy transitions" and the PPCA, for example. Critics note, however, that loopholes in such policies still allow some of its partners and intermediaries to fund some coal projects (Willis and Indri 2023). In the decade following the World Bank's declared moratorium, the multilateral and private finance sector changed substantially where coal financing is concerned. A 2023 report by the Institute for Energy Economics and Financial Analysis (Trivedi and Srivastava 2023: 4) found that "over 200 significant financial institutions have established coal exclusion policies, with divestment momentum away from coal accelerating in the last two years." This number includes eighty-seven banks, fifty-one insurers and reinsurers, thirty-six asset management firms, and twenty-seven export credit agencies, multilateral development banks, and development finance institutions. While just over half of the institutions are based in Europe, the others are from the Asia-Pacific region, North America, Africa, and Latin America. Almost a quarter strengthened their coal export policies in 2021 or 2022. The report is filled with evidence and language about the growing "momentum" in favor of coal exclusion policies.

However, 2024 brought evidence of some major financial firms "flip-flopping" or "retreating" from their coal exclusion and other climate change action commitments (Gelles 2024; Sorkin et al. 2024). A combination of political backlash against such commitments in many US states and the Republican Party and concerns about investor lawsuits related to such firms' fiduciary responsibilities seems to be driving these changes among some North American financial services corporations. Also of concern to climate activists and researchers are the various ways that financial institutions continue to help raise money for coal, oil, and gas investments even after announcing restrictive policies – as happens when they participate in corporate bond markets, for example. One report estimated that European banks, most with restrictive fossil fuel policies related to their direct investments, helped fossil fuel firms raise more than EUR 1 trillion on such bond markets from 2016 to 2023 (Ambrose 2023) – that is, *after* the 2015 UNFCCC climate summit in Paris.

The flip-flopping recently seen among financial institutions' coal-specific and/or broader fossil fuel policies serves as a clear reminder that normative change rarely occurs in a simple, singular, linear, or inevitable path. The more long-standing set of energy policy debates that illustrates this point can be seen around nuclear energy phaseout policies, where Germany and Sweden have played host to several flip-flops in policy and/or controversial delays in phaseout implementation (Johnson 2023; Maquire 2023). Subsequent governments may bring different views into power, or subsequent events (like energy price spikes or war) can certainly result in phaseout schedules or commitments – or existing bans being reversed. In Sweden, a new conservative government (after 2022) seeks to reverse the nuclear phaseout planned since the 1980s, while in Germany, national leaders have struggled and sometimes delayed various plans and implementation goals as they seek to implement planned phaseouts of both coal burning and nuclear power.

5.5 Politicizing and Contesting Phaseouts and Financing Bans

As coal phaseout norms have proliferated across states, multilateral and transnational political fora, and multilateral and multinational financial institutions, a host of actors and arguments that critique, contest, and oppose coal phaseout policy norms have emerged in discourses and political action from local to global scale. Objections to and grievances about coal phaseout policies commonly invoke justice and injustice – particularly three forms of justice common in climate change politics and research: distributional, procedural, and recognitional (Newell et al. 2021; Zimm et al. 2024).

Local- and national-scale critics of coal phaseout policies commonly embed their concerns in distributional justice concerns, focusing of the costs borne by coal sector workers, families, local communities, and coal-dependent regions and the need for substantial compensation and interventionist policies (Bang et al. 2022; Busch et al. 2023; Jakob et al. 2020; Steckel and Jakob 2022; Wong et al. 2022). Such politics can be easily identified at the local and state/provincial levels in coal mining regions of the United States, Spain, and Germany, for example, but research demonstrates such justice-related concerns and debates are quite similar across the Global North/South binary (Busch et al. 2023).[6] Because more coal phaseouts have been achieved in parts of the Global North, this literature draws heavily on those cases. Nevertheless, coal-dependent regions of states like China, India, South Africa, and Indonesia seem likely to face similar challenges if national and global governance increasingly prioritizes decarbonization policies. In many respects, how these objections are articulated and addressed (or mostly ignored) in

[6] See also Bernstein and Hoffmann's discussion of just transitions, this volume.

provincial and national politics might be expected to mirror comparative politics dynamics and research around other issues combining domestic material interests and moral arguments/imperatives.

What levels of support for displaced workers and local and regional economic damages should be borne by the larger society via state and non-state organizations? Few advocates for local- and national-scale justice want to emulate the violence and economic and social injustices meted out by the UK's Thatcher government on miners and mining families, trade unions, communities, and regions. Drawing lessons from this era, Paterson's (2024) challenging question asks, "can you destroy the power base for fossil fuels without creating massive injustice?" The economic, ecological, social, cultural, and political legacies of many former UK and US coal mining regions pose a host of complex justice questions for such communities around the world if coal is to be rapidly phased out, and fossil gas and oil are soon to follow. It should be noted, however, that the ecological and human and community health costs of continued coal burning and other fossil fuel infrastructures are also rife with distributional justice concerns, including highly racialized fossil fuel burdens in the United States (Donaghy et al. 2023).

Global- or transnational-scale critiques of coal phaseout policies invoke a complex mix of procedural and distributional concerns. Procedural justice concerns are often related to the principle of sovereign authority and/or to contemporary decolonial discourses and critiques (Biddau et al. 2024; Brown and Speigel 2019; Feffer 2023; Hamouchene 2022; Malm 2016b; van Ryneveld and Islar 2023). Among the paradoxes in these debates and scholarship is that colonial power often drove substantial portions of fossil fuel system and construction across parts of the Global South, while contemporary concerns about growing "green" or renewable energy "colonialism" are increasingly expressed. The use of power and influence by European and North American states in both bilateral relations and multilateral fora increasingly appears as a chief concern, whether such influence is being exercised in UNFCCC processes (often by states who have yet to phase out coal and/or have rather poor decarbonization records at home) or in multilateral development banks and private sector financial institutions. In other words, the growing pressure – or "incentives" – for states in the Global South to phase out coal increasingly engenders all of the same concerns associated with the exercise of power in many areas of global governance and international organizations. Such concerns often note the substantial fraction of historical GHG emissions sourced in the Global North and the generally quite slow pace of such countries' domestic decarbonization at home.

An additional critique focuses on the aggregate amount of financial assistance available, whether one looks to current and historical pledges made within UNFCCC processes or the total amount of financial assistance available in various multilateral and private financial institutions for the alleviation of energy poverty and the expansion of reliable and adequate electricity in the Global South. Are states and societies in India and South Africa expected to bear most or all of the costs of slowing, reducing, and eventually ending coal mining and coal burning?

5.6 Whither Coal and Politicized Fossil Fuel Infrastructures?

To conclude, I focus discussion on four sets of implications of contemporary coal phaseout politics: for advocates of more substantial climate change mitigation policies; for skeptics of many aspects of globalized prohibitions in climate change politics; for the stability and politicization debates at the center of this volume; and for the potential benefits of more systematic comparative politics within globally and transnationally framed research.

For climate change mitigation policy advocates, the "success" (from their perspective) of coal phaseout policy and goal proliferation – and of increasing implementation of many of those policies – serves as a model for discursive and material strategies aimed at other fossil fuel decarbonization goals. These goals include phaseouts for internal combustion engines, for diesel trains/rail, for the expansion of natural gas infrastructures, for the reduction and phasing out of natural gas burning for electricity production, and for more general anti-fossil goals/norms. In the medium term, coal phaseout activists seek to increasingly isolate and target the small number of states which now constitute the vast majority of global coal burning. More recent climate change-related prohibition movements focusing on a coal elimination treaty (Burke and Fishel 2020), leaving fossil fuels in the ground and advancing anti-fossil fuel norms (Newell and Simms 2020), and efforts seeking geoengineering non-use agreements (Biermann et al. 2022; Gupta et al. 2024; VanDeveer et al. 2024) all seek to explicitly politicize areas of global climate policy and governance as a means to establish and promulgate climate change-related global prohibition regimes.

For coal phaseout skeptics and opponents, the political dynamics explored here suggest the continued use of combinations of material and discursive power to shape anti-fossil fuel politics generally. For such actors, (re)politicizing anti-fossil fuels norms and their implementation via existing organizations and material constraints seems likely to focus on distributional, procedural, and distributive justice concerns. These actors will remain reluctant to prioritize climate change above issues of local and global poverty alleviation, representational and procedural priorities within global governance, and a host of concrete outcomes related to the need to mitigate and eliminate energy poverty, food insecurity, and other areas of human insecurity. In this way, climate change and energy policies will continue to be critiqued as an "arena" of global-, national-, and local-scale policymaking that must focus on economic, social, and global justice and the institution of transnational power and influence – not solely or primarily on climate change per se.

Coal phaseout politics teach us about the antagonisms between stability and (re) politicization. First, while coal has long been framed in highly politicized ways in national and international politics and policymaking, it takes concerted and sustained effort by diverse sets of agents to politicize burning it and to reframe politics around phasing out its use entirely – and banning the financing and burning of coal for electricity generation. Such agents have, in many domestic, transnational, and interstate fora, critiqued and politicized burning it at all, as well as related issues like financing coal infrastructures of all kinds and mining, transporting, and exporting it. Interestingly, it seems likely that some depoliticized developments and discourses, such as those associated with the price of renewable energy and the co-benefits of coal phaseouts, afforded opportunities to politicize coal burning.

In seeking to phase out and permanently ban coal burning, many advocates seek to move from one stable set of norms and policies to another. But this is not a case of simply moving from one stable set of policies and associated expectations to another stable state, via a short period of politicizing policy, governance, and outcomes. Instead, phaseout policies have been the subject of substantial local-scale critique in coal mining communities and coal mining national polities, as well as the subject of explicitly normative and ideological critique of "imposing" coal phaseouts across North/South or colonial/decolonial lines. Stabilizing new policy norms associated with phasing out and permanently banning coal burning attracts the attention of agents who object to the new norms by invoking both distributive and procedural justice arguments. In fact, the cross-scale politics of justice may invite more complex, iterated politicization, given the essentially contested nature of various forms of justice. The implication here being that many agents can make and deploy different and conflicting claims of justice and injustice.

Lastly, for comparative and international environmental, climate, and energy politics scholarship, it also suggests the need for a more systematic comparative politics of coal mining, investment, exports, imports, and coal phaseouts. What domestic and transnational factors make coal phaseout policies more/less likely to be enacted? And what causal factors result in effective implementation, or political backlash or rollback of such policies? Such research can draw on the growing comparative environmental politics scholarship (Sowers et al. 2023; Steinberg and VanDeveer 2012) – as can work on the comparative politics of climate change and green energy backlash and policy rollbacks. So far, there appear to be few examples of state rollbacks of coal phaseout policies. As noted in this chapter, however, that is not true of private sector financial institutions. As similar "phasedown" and/or "phaseout" political demands and policy dynamics develop around natural gas – whether or not to continue to invest in expanded gas infrastructures of all kinds or whether to pledge phasedowns and/or phaseouts by particular dates – what is needed is a more robust comparative politics research agenda examining anti-fossil fuel activism and decarbonization policymaking more broadly across states and other jurisdictions, and across civil society and private sector organizations. The politics and governance of energy transitions and decarbonization pathways seem quite unlikely to produce stable, linear, and uncontested policies over time and across scales, jurisdictions, and sectors.

References

Adekoya, O. B., Kenku, O. T., Oliyide, J. A. and Al-Faryan, M. A. S. (2023). On COP26 and coal's phase-out agenda: Striking a balance among environmental, economic, and health impacts of coal consumption. *Journal of Environmental Management* 328, 1–11.

Ambrose, J. (2023). Europe's banks helped fossil fuel firms raise more than 1 trillion Euro from global bond markets. *The Guardian*, September 26.

Anderson, K., Broderick, J. and Stoddard, I. (2020). A factor of two: How mitigation plans of "climate progressive" nations fall short of Paris-compliant pathways. *Climate Policy* 20(10), 1290–1304.

Bang, G., Rosendahl, K. E. and Borhringer, C. (2022). Balancing cost and justice concerns in the energy transition: Comparing coal phase-out policies in the UK and Germany. *Climate Policy* 22(8), 1000–1015.

Biddau, F., Rizzoli, V. and Sarrica, M. (2024). Phasing out "coal tradition" in favour of "renewable colonialism": How the press contributes to the discursive (de)legitimization of coal and renewables in a coal region in transition. *Sustainability Science* 19, 381–402.

Biermann, F., Oomen, J., Gupta, A. et al. (2022). Solar geoengineering: The case for an international non-use agreement. *WIREs Climate Change* 13(3), e754.

Blondeel, M., Colgan, J. and Van de Graaf, T. (2019). What drives norm success? Evidence from anti-fossil fuel campaigns. *Global Environmental Politics* 19(4), 63–83.

Blondeel, M., Van de Graaf, T. and Haesebrouck, T. (2020). Moving beyond coal: Exploring and explaining the powering past coal alliance. *Energy Research & Social Science* 59, 101304.

Boersma, T. and VanDeveer, S. D (2016a). Coal after the Paris Agreement. *Foreign Affairs*, June 6.

Boersma, T. and VanDeveer, S. D. (2016b). The G20 after Paris: Curbing carbon emissions. In G20 Research Group, ed., *G20: The Hangzhou Summit*, p. 19.

Brown, B. and Spiegel, S. J. (2019). Coal, climate justice and the cultural politics of energy transition. *Global Environmental Politics* 19(2), 149–168.

Burke, A. and Fishel S. (2020). A coal elimination treaty 2030: Fast-tracking climate change, global health and security. *Earth System Governance* 3, 100046.

Busch, H., Ramasar, V., Avila, S. et al. (2023). Mining coal while digging for justice: Investigating justice claims against a coal-phase out in five countries. *The Extractive Industries and Society* 15, 101275.

de Moor, J. (2023). Time and place in climate activism: Three urgency-induced debates. In J. Sowers, S. D. VanDeveer and E. Weinthal, eds., *Oxford Handbook of Comparative Environmental Politics*. Oxford: Oxford University Press, pp. 299–316.

Dimitrov, R. (2016). The Paris Agreement on climate change: Behind closed doors. *Global Environmental Politics* 16(3), 1–11.

Do, T. N. and Burke, P. J. (2023). Phasing out coal power in a developing country context: Insights from Vietnam. *Energy Policy* 176, 113512.

Donaghy, T. Q., Healy, N., Jiang, C. J. and Battle, C. P. (2023). Fossil fuel racism in the United States: How phasing out coal, oil, and gas can protect communities. *Energy Research and Social Science* 100. https://doi.org/10.1016/j.erss.2023.103104

Ember (2024). *European Electricity Review 2024: Europe's Electricity Transition Takes Crucial Strides Forward.* London: Ember. https://ember-energy.org/app/uploads/2024/10/European-Electricity-Review-2024.pdf

Feffer, J. (2023). Don't let the green energy transition become a new colonialism. *Fair Observer*, October 23.

Finnemore, M. (1996). *National Interests in International Society*. Ithaca, NY: Cornell University Press.

Freese, B. (2003). *Coal: A Human History*. New York: Basic Books.

Friedlingstein, P., O'sullivan, M., Jones, M. W. et al. (2022). Global carbon budget 2022. *Earth System Science Data Discussions* 14(1), 4811–4900.

Gelles, D. (2024). More Wall Street firms are flip-flopping on climate. *New York Times*, February 19.

Green, F. (2018a). Anti-fossil fuel norms. *Climatic Change* 150, 103–116.

Green, F. (2018b). The logic of fossil fuel bans. *Nature Climate Change* 8, 449–451.

Gupta, A., Biermann, F., van Driel, E. et al. (2024). Towards a non-use regime on solar geoengineering: Lessons from international law and governance. *Transnational Environmental Law* 13(2), 368–399.

Hamouchene, H. (2022). The energy transition in North Africa: Neocolonialism again! Transnational Institute, October 14. www.tni.org/en/article/the-energy-transition-in-north-africa

Jaeger, J. (2023). These 10 countries are phasing out coal the fastest. World Resources Institute, November 30. www.wri.org/insights/countries-phasing-out-coal-power-fastest

Jakob, M. and Steckel, J. C., eds. (2022). *The Political Economy of Coal: Obstacles to Clean Energy Transitions*. Abingdon: Routledge.

Jakob, M., Steckel, J. C., Jotzo, F. et al. (2020). The future of coal in a carbon-constrained climate. *Nature Climate Change* 10(8), 704–707.

Johnson, S. (2023). Sweden plans new nuclear reactors by 2035, will share costs. *Reuters*, November 16. www.reuters.com/business/energy/sweden-plans-new-nuclear-reactors-by-2035-can-take-costs-2023-11-16/

Karlsson, M. and Gilek, M. (2020). Mind the gap: Coping with delay in environmental governance. *Ambio* 49, 1067–1075.

Malm, A. (2016a). *Fossil Capital: The Rise of Steam Power and the Roots of Global Warming*. London: Verso.

Malm, A. (2016b). Who lit this fire? Approaching the history of the fossil economy. *Critical Historical Studies* 3(2), 215–248.

March, J. and Olsen, J. (1989). The institutional dynamics of international political orders. *International Organization* 52(4), 943–969.

Medzhidova, D. (2022). Return of coal: A short visit or a long stay? *BRICS Journal of Economics* 3(4), 209–229.

Merry, H. J. (1955). The European Coal and Steel Community: Operations of the high authority. *Western Political Quarterly* 8(2), 166–185.

Misik, M. and Pracharova, V. J. (2023). Coal is a priority for energy security, until it's not. *Sustainability* 15(8), 6879.

Mitchell, R. B. and Carpenter, C. (2019). Norms for the Earth: Changing the climate on "climate change." *Journal of Global Security Studies* 4(4), 413–429.

Mitchell, T. (2013). *Carbon Democracy: Political Power in the Age of Oil*. London: Verso.

Muttitt, G., Price, J., Pye, S. and Welsby D. (2023). Socio-political feasibility of coal power phase-out and its role in mitigation pathways. *Nature Climate Change* 13(February), 140–147.

Nadelman, E. A. (1990). Global prohibition regimes: The evolution of norms in international society. *International Organization* 44(4), 479–562.

Newell, P. and Simms, A. (2020). Towards a fossil fuel non-proliferation treaty. *Climate Policy* 20(8), 1043–1054.

Newell, P., Srivastava, S., Naess, L. O., Contreras, G. A. T. and Price, R. (2021). Toward transformative climate justice: An emerging research agenda. *WIREs Climate Change* 12(6), e733.

O'Neill, K., Balsiger, J. and VanDeveer, S. D. (2004). Actors, norms and impact: Recent international cooperation theory and the influence of the agent-structure debate. *Annual Review of Political Science* 7, 149–175.

Ohlendorf, N., Jakob, M. and Steckel, J. C. (2022). The political economy of coal phase-out: Exploring actors, objectives, and contextual factors shaping policies in eight major coal countries. *Energy Research & Social Science* 90, 102590.

Paterson, M. (2024). An unjust transition. *Ecologist*, February 24.

Paterson, M., Tobin, P. and VanDeveer, S. D. (2022). Climate governance antagonisms: Policy stability and repoliticization. *Global Environmental Politics* 22(2), 1–11.

Paxman, J. (2022). *Black Gold: The History of How Coal Made Britain*. London: William Collins.

Rentier, G., Lelieveldt, H. and Kramer, G. J. (2019). Varieties of coal-fired power phase-out across Europe. *Energy Policy* 132, 620–632.

Selin, H. and VanDeveer, S. D. (2025). Global climate change governance: Where next, after Paris? In M. Kraft, B. Rabe, and N. J. Vig, eds., *Environmental Policy* (12th ed.). Thousand Oaks, CA: SAGE/CQ Press, pp. 249–267.

Sorkin, A. R. et al. (2024). "Wall Street's Climate Retreat." *New York Times*, February 16.

Sowers, J., VanDeveer, S. D. and Weinthal, E., eds. (2023). *Oxford Handbook of Comparative Environmental Politics*. Oxford: Oxford University Press.

Spasic, V. (2024). Coal plants in Bulgaria to be shut not by government but by market. Balkan Green Energy News, January 15.

Steckel, J. and Jakob, M. (2022). To end coal, adapt to regional realities. *Economist* 607(7), 29–31.

Steinberg, P. and VanDeveer, S. D. (2012). *Comparative Environmental Politics: Theory, Practice and Prospects*. Cambridge, MA: MIT Press.

Stoddard, I., Anderson, K., Capstick, S. et al. (2021). Three decades of climate mitigation: Why haven't we bent the global emissions curve? *Annual Review of Environment and Resources* 46, 653–689.

Trivedi, S. and Srivastrava, S. (2023). 200 and counting: Global financial institutions are exiting coal. Institute for Energy Economics and Financial Analysis, May 4.

Twidale, S. (2024). Global energy-related CO_2 emissions hit record high in 2023, *Reuters*, March 1.

US EIA (United States Energy Information Agency) (2024). Global CO_2 emission rise through 2025 in most IEO 2023 cases. *Today in Energy*, November 20.

van Asselt, H. and Green, F. (2023). COP26 and the dynamics of anti-fossil fuel norms. *Wiley Interdisciplinary Reviews: Climate Change* 14(3), e816.

van Ryneveld, T. N. and Islar, M. (2023). Coloniality as a barrier to climate action: Hierachies of power in a coal-based economy. *Antipode* 55(3), 958–981.

VanDeveer, S. D. (1997). Normative force: The state, transnational norms and international environmental regimes. PhD diss., University of Maryland.

VanDeveer, S. D. and Boersma, T. (2022). World on fire: Coal politics and great power responsibility. In R. Falkner and B. Buzan, eds., *Great Powers, Climate Change and Global Environmental Responsibilities*. Oxford: Oxford University Press, pp. 248–276.

VanDeveer, S. D., Biermann, F., Kim, R. E., Bardi, C. and Gupta, A. (2024). Three pathways to nonuse agreement(s) on solar geoengineering. *Ethics & International Affairs* 38(3), 275–286.

Vinichenko, V., Vetier, M., Jewell, J., Nacke, L. and Cherp, A. (2023). Phasing out coal for a 2C target requires replication of the most ambitious national plans despite security and fairs concerns. *Environmental Research Letters* 18(1). https://doi.org/10.1088/1748-9326/acadf6.

Willis, D. and Indri, N. (2023). Coal not yet confined to the "old days" by World Bank Group. Bretton Woods Project, December 13. www.brettonwoodsproject.org/2023/12/coal-not-yet-not-confined-to-the-old-days-by-world-bank-group/.

Wong, J., Roser, F. and Maxwell, V. (2022). *Coal Phase-Out and Just Transitions: Lessons from Europe*. Cologne: NewClimate Institute.

Zimm, C., Mintz-Woo, K., Brutschin, E. et al. (2024). Justice considerations in climate research. *Nature Climate Change* 14(1), 22–30.

6

Peaty Politics

MATTHEW PATERSON

This chapter explores a particular sort of tension between the pursuit of policy stability and the recurrent politicization of climate change. As climate policy regimes, focused on engineering a smooth pathway to shift emissions toward (net) zero over time, develop, they bring new objects of governance into view. These objects have their own preexisting political fields and dynamics. When policy regimes on climate change were established, the initial focus was for the most part on the energy sector. This has been for good and obvious reasons – fossil energy remains comfortably the single most important source of climate change, accounting for 65–75 percent of overall greenhouse gas emissions in 2010 (fossil carbon dioxide is 65 percent plus, while methane is 16 percent, a substantial amount of which comes from fossil fuels; see IPCC 2014: 7). As a result, policy was focused on shifting supply in the electricity sector, pursuing energy efficiency, and once electricity decarbonization was underway, electrifying direct fossil fuel use. At the same time, the political problems of this approach are well-known: We have many analyses of the way that incumbent interests in fossil fuels (oil, coal, gas) and their key allied sectors (aviation, automobiles, fossil fuel-dependent manufacturing like cement, steel, and plastics) have been able to turn the pursuit of stable emissions reductions into stability as the status quo (e.g. Brulle 2014; Newell and Paterson 1998; Stokes 2020).

But within the logic of a stable policy regime is a logic of what I call *scope expansion* in climate policy. The more or less exclusive focus on the energy sector made sense in a world where emissions only needed to be reduced by 60 percent, as the first Intergovernmental Panel on Climate Change (IPCC) report stated in 1990. But as the emissions reductions needed to respond adequately to the climate crisis have increased, so has the scope of climate policy: Remaining sources of emissions have become the object of attention for both researchers and governments, and carbon sinks also come into view in a deeper way. This process generates new sorts of potential for political conflict as new actors and power relations are brought into play in relation to climate change policy. Some of this focus regards the so-called hard-to-abate sectors such as steel and cement (Bulkeley et al. 2022). But land use, agriculture, and food systems have also become increasingly the object of attention. In the Global South, land use emissions have long been much more important, and deforestation has been highly important to both the emissions profile of many countries (most famously Brazil; see Hochstetler, Chapter 9, this volume) and the contentious politics of climate change. The shift in attention is thus to land use questions in the Global North, triggered by the need for such countries to aim at "net zero" emissions by around 2050 at the latest, to meet the goals of the 2015 Paris Agreement.

This chapter turns its attention to one aspect of scope expansion in climate policy action into land use in the Global North: the focus on peatlands in the UK. Peatlands are global in scope, and there are wider initiatives such as the Global Peatlands Initiative, which hosted a large pavilion at the 26th United Nations Framework Convention on Climate Change (UNFCCC) Conference of the Parties (COP26) in Glasgow along with various other partners (Global Peatlands Initiative 2021). Peat has also become an object of climate governance in other countries (on Norway, see Farstad et al. 2022; on Finland, see Ratamäki et al. 2019; on Ireland, see Carter and Little 2021). As such, we see an emerging comparative politics of peat. But peatlands are a particularly prominent part of the land use and climate change picture in the UK, and a focus on the UK demonstrates particularly well these dynamics of scope expansion bringing new actors and conflicts into view. The UK is also a useful case in this regard precisely because it is at the forefront of such scope expansion – having eliminated coal from electricity production and being well advanced in decarbonizing electricity, it is now in the position of having to move to other sites to pursue climate ambition beyond electricity. It is thus instructive for the challenges that many other countries may face in the near future.

The dynamics of stability and politicization still play out in this case, but they do so rather differently. Peatlands, especially in mountain areas, are far from central to contemporary economies overall, compared to fossil energy. The power of those who control such spaces therefore does not derive from the structural qualities that fossil energy has. But they are important for a range of other reasons. A focus on this politics of peat emphasizes one element in the book's overall narrative: that the pursuit of stability in terms of either long-term emissions pathways or policy design over time often ends up reproducing stability as the status quo (see Paterson, Tobin, and VanDeveer, Chapter 1, this volume).

6.1 Peat as an Object of Climate Governance

Peat emerged as an important element in climate politics and governance because of its particularly important role in land-based carbon sequestration. Peatlands are the biggest single terrestrial store of carbon globally (Joosten et al. 2016) and thus crucial to the climate challenge (Gewin 2020). Yet they are simultaneously threatened through forestry, agriculture, and economic development initiatives (Ewert and Abel 2021). The UK is host to some of the most important peatlands, and they play a particularly important role in the potential for land-based carbon sequestration in the UK, alongside afforestation. Peatlands account for around 8 percent of the UK land area (IUCN 2018). This is even more so the case in Scotland, where a disproportionate amount of the peatlands in the UK are located and whose land-cover is approximately 20 percent peatland (Scottish Government 2015). These can be usefully divided into upland and lowland peatlands.[1] The upland areas would be termed in general peat moors, and when they are well-functioning ecosystems they would be called blanket bogs, emphasizing the wet character of such ecosystems. I focus mostly on peat moors in this chapter, partly for reasons of space but partly because the complex dynamics of politicization around peat moors demonstrates well the limits of technocratic policymaking.[2]

The UK's climate policy architecture is often taken as the paradigm of a system focused on long-term policy stability: The 2008 Climate Change Act sets a long-term goal (originally 80 percent cuts by 2050, updated to "net zero" in 2019) and requires the UK government to set five-year budgets significantly in advance that must be consistent with the pursuit of that long-term goal. The act also created a Climate Change Committee (CCC), an arms-length expert agency with the responsibility for both evaluating and assessing the government's progress toward the long-term goal and providing critical policy input to generate new initiatives to keep the pursuit of that goal on track. As such, it is often lauded as one of the most ambitious and elaborate climate policy architectures globally (Dubash et al. 2021; Fankhauser et al. 2018; Lockwood 2021), and certainly was the result of considerable political contestation (Carter 2014; Carter and Jacobs 2014). But it is also the paradigm of a depoliticized institutional arrangement, seeking to insulate the pursuit of the overall goal from political contestation, and channel the latter where it exists into either the politics of the design of specific policies or the politics of performance over whether governments have adequate measures in place to achieve the goal (Kuzemko 2016; Lockwood 2013). This depoliticization of

[1] Lowland peatlands are usually also divided into raised bogs, where peat accumulates and creates a raised wet area compared to surrounding land, and fenlands where the water is much more dominant and the peat submerged.

[2] It is worth noting briefly how this shapes the specifics of the argument I develop. Lowland peat areas are often of very considerable agricultural value, since peat is such a rich form of soil. They (and especially fenlands) have been extensively drained for agricultural use since the Middle Ages as a result. They also have been used commercially through peat cutting for horticultural compost, which has become a source of conflict and I discuss briefly in Section 6.2.1. (In some areas, notably northwest Scotland, they are also still cut for domestic fuel use, as is the case much more famously in Ireland.) The landownership issues discussed in this chapter thus operate rather differently in lowland areas than on the peat moors.

climate strategy has been largely successful, and the architecture has largely survived major political shocks such as Brexit (Farstad et al. 2018). It has also, at least until very recently, so far survived contestations from within the Conservative Party while it has been the governing party (Carter and Pearson 2024). This opposition to climate policy has become more vociferous from 2021 onward with the establishment of the Net Zero Scrutiny Group (Paterson, Wilshire, and Tobin 2023). At the time of writing (September 2023) the government under Rishi Sunak has been taking more significant steps to roll back key parts of UK climate policy (e.g. Jackson 2023), raising important questions about the durability of the UK climate policy architecture exhibits in the face of such a shift.

As the UK's climate policy architecture developed, and in particular the overall target shifted first toward 80 percent cuts (in the 2008 Climate Change Act target) and then net zero from 2019, the focus on peatlands became progressively more important to its delivery. This is because of the increased centrality of carbon sequestration: Peatlands have the highest carbon density of any land ecosystem (IUCN 2018: 8) and thus their potential for carbon sequestration is particularly strong. At the same time, around 80 percent of the UK's peatlands are highly degraded and net sources of carbon, rather than sinks of it, as well as considerably lacking in biodiversity (IUCN 2018).

The centrality of peatlands to land-based sequestration was highlighted in the UK government's CCC report *Land Use: Policies for a Net Zero UK* (CCC 2020). Peatlands also have become important because of the way (like afforestation at least in some areas) that they enable the intertwining of mitigation and adaptation – restored peatlands will have significant beneficial effects on flood management, dramatically reducing the rate of water runoff in heavy rains (Goudarzi et al. 2021; Shuttleworth et al. 2019; Vinter 2021). This quality is particularly important in the UK, where enhanced flooding is a key climate impact already being experienced as rainfall patterns change. The *Land Use: Policies for a Net Zero UK* report also suggests that strategies to restore peatlands are a highly significant aspect of pursuing net zero, with the potential to cut emissions from land use broadly on a similar scale to forestry, and on a per hectare basis, considerably more so. Emissions from peatlands themselves currently account for about 1 percent of UK emissions (with the percentage of Scotland's being correspondingly a good deal higher), so they are small, but there is potential not only to eliminate these emissions through proper peatland management but to reverse this and make peatlands store large amounts of carbon rather than release it.

Both the UK and Scottish governments have established targets and initiatives to enable peatlands to play this role in pursuing net zero. The UK-wide strategy, established in 2013, is managed by the international scientific environmental nongovernmental organization (ENGO) the International Union for Conservation of

Nature (IUCN) (IUCN 2018). It aims to have 2 million hectares of peatland in "good condition, under restoration or being sustainably managed" by 2040 (IUCN 2018: 12). The Scottish government's climate change plan 2018–2032 aims to restore 250,000 hectares (about 3 percent of Scotland's land area) of degraded peatland by 2030 (Scottish Government 2015, 2020).

These government strategies to pursue peatland restoration operate via government–NGO–landowner partnerships and through grant funding programs by government, in particular the Nature for Climate Peatland Grant Scheme (Natural England 2021). Public funding has generated widespread initiatives and partnerships for peat restoration, including, for example, Moors for the Future, the Northumberland, Yorkshire, Lancashire peatland partnerships, and the "Great North Bog." These are usually partnerships between IUCN, local wildlife trusts, and individual landowners (including large NGO landowners like the National Trust but also private landowners). They engage in highly technical projects to restore peatlands. At their core, such projects mostly entail rewetting the bogs by preventing water runoff (in mountain bogs in particular) to keep them wet, and then reintroducing plant species (sphagnum moss is the central species) to sustain these rewetted bogs and prevent them drying out (Shuttleworth et al. 2019). Once the moors are rewetted (to become bogs), they then become sinks for carbon rather than sources (as the peat dries out it becomes highly friable and blows away readily), as well as functioning as highly diverse and distinctive ecosystems (home, for example, to the UK's only carnivorous plant, the sundew).

Alongside these government-funded initiatives, there is a parallel development of privately funded peatland initiatives, alongside those for afforestation, which operate on a logic of carbon offsetting by large corporations seeking to offset emissions to present themselves as "net zero," or private individuals seeking to develop projects to sell carbon offset credits on to such corporate actors, as part of their income stream. These private initiatives are often the target of significant criticism for their "land grab" qualities, on which see more in Section 6.2.2 (Garavelli 2022; Macfarlane 2021).

Given that it is embedded in the broader architecture of the UK climate governance system, this approach to peat reflects the desire for long-term stable policy environments to pursue decarbonization: The quantities of carbon to be stored by peatland restoration initiatives all figure in the carbon budget calculations over time to meet the UK's long-term targets. The various partnerships and initiatives represent well a depoliticized strategy of focusing on the technical qualities of the peat restoration initiatives and the broad social partnerships between the different actors through which they are organized. These partnerships and initiatives exemplify Harriet Bulkeley's (2016) notion of climate governance being something that is about "accomplishing" – it is patient, slow, experimental. But it also then entails an assumption – usually implicit – that there are no underlying conflicts; it is just about patiently working on both the peatlands and the partnerships to pursue climate governance.

6.2 The Contested Politics of Peat

But peat politics focused on rewetting initiatives sits uneasily alongside the other political qualities of peat that have been brought into view by UK climate policy "scope expansion," and which shape the possibilities for peat to play a full role in the overall response to climate change. Three in particular are important to draw out in this context: how peatlands have been important sites of social movement campaigning and struggle; the embeddedness of peat ownership and management in the UK class structure; and peat moors as sites of aestheticized leisure consumption.

6.2.1 Social Struggles over Peatlands

The first such political quality is that peat operates as a site of social movement campaigning and struggle. A common focus of such campaigns, rising periodically for a few decades, is over the peat in garden compost. Lowland peat from fens and lowland bogs is extracted and used widely as a base for compost, given its rich composition of organic matter. The UK government has a current proposal to ban peat in commercial garden compost by 2024 (UK Government 2022). This issue is mostly about lowland peat where peat is extracted for commercial compost use, rather than upland moors.

On peat moors, recurrent campaigning activity has been over two distinct areas. First, there are currently widespread local campaigns to save upland peat moors focused on the question of winter burning and, by extension, grouse shooting. A common focus of campaigns is on the winter burning of the heather that is the dominant plant species on the (dried-out) peat moors. The heather is burned to stimulate new growth of young heather shoots, which are favored by the grouse and thus maximize the yield of grouse for the annual shoots in late summer and early autumn. From the point of view of landowners, this process thus maintains an ecosystem most suitable for the red grouse. The winter burning, however, is highly controversial. Apart from the immediate carbon (and other) emissions both from the heather itself and from the underlying peaty soil, winter burning maintains the peat in a dry condition, which releases carbon directly and creates additional fire risks in the summer. It also generates enhanced flood risks as the lack of a rich and wet plant ecosystem accelerates water runoff from the moors, often into narrow, densely populated valleys of northern England. Additionally, burning means that the potential of the peatlands for absorbing carbon is eliminated, and that an environment with low biodiversity centered on grouse is maintained. The burning is regulated to limit the amount of peat loss and the risk of fires spreading, and these regulations were tightened in 2021 to require licenses if a landowner wanted to burn plants on peat deeper than a depth of 40 cm, but there is considerable evidence of illegal burns by landowners and managers (Horton 2022b; Rowlatt 2022; RSPB 2022), including directly in the run-up to COP26 in late 2021 (Richards 2021).

Landowners are vehemently in favor of winter moorland burning. Their lobby group, the Moorland Association, threatened legal action in February 2020 when the government discussed banning moorland winter burning (Evans 2020). They use various arguments to defend peat burning as a practice. These include suggesting that there is no real alternative and if they didn't do this the moors would, in the words of Amanda Anderson, director of the Moorland Association, then be degraded by "afforestation, windfarms or overgrazing" (as quoted in Carrington 2016). Elsewhere, they argue that the claims for sphagnum moss projects like Moors for the Future's about carbon sequestration are "somewhat unproven" (Game and Wildlife Conservation Trust 2014).

The second sort of campaigning over peat moors has centered on questions of access to land for walking and other leisure activities. While the enclosure of land in England from the fifteenth century onward never fully eliminated older commoners' rights to land, landowners progressively tried during the nineteenth and early twentieth centuries (and since in some notable instances) to prevent access to land for walking that people previously had rights to. From the late nineteenth century onward, the newly organized urban working classes in northern England in particular led the charge for land access through a series of mass trespasses – large numbers of people walking on private land to claim rights to walk, which frequently culminated in their being beaten by landowners' agents or the police and arrested and imprisoned for their actions (Glasby 2012; Hayes 2021; Hill 1980). The most famous of these was the Kinder Scout trespass of 1932, on Kinder Scout in Derbyshire, which is a large peat moor at the southern end of the Pennines that range from the English Midlands north across the Scottish border. This trespass is usually credited with having led to legislation in the immediate post-World War II period to formalize public rights of way in England and Wales (e.g. Glasby 2012; but cf. Hey 2011), even if the resolution of these conflicts never matched campaigners' ambitions for full "rights to roam" (Shoard 1987; Shrubsole 2019: 250–252). The widespread access this gives (not as widespread as in many European countries but radically more than in North America) and the set of leisure pursuits, especially hillwalking, that it enables have then fed into social support for campaigning around peat moor management.

These conflicts over access to land, where upland peat moors remain central sites of contention, are ongoing. In the late 1990s, the Labour government introduced legislation to enhance the "right to roam" in England and Wales, but landowners managed to curtail that legislation considerably (Parker 2008). The right to roam only covers 8 percent of England's land area (Horton 2022a), and after a review during 2021–2022, the government announced it would not expand this area, defending the decision as "the countryside is not just a place of leisure, but it is also a place of business" (as quoted in Horton 2022a).

6.2.2 Peat and the UK's Class Structure

Underlying these political conflicts over access and grouse moor management is the way that peat moors are embedded in the social structure of the UK. Reflecting broader patterns in the UK (Christophers 2018), land ownership on peat moors is highly concentrated; indeed, given it cannot operate as agricultural land in any real sense and it is largely open unenclosed land, it makes little sense as small holdings. Some of this land is owned by public authorities (e.g. the UK Ministry of Defence, which owns 1.6 percent of UK land, mostly for army training; see Ministry of Defence 2022) and some by large charities (principally the National Trust, which now owns most of the top of Kinder Scout, for example). But most is privately owned, and by a relatively small number of people. In Scotland, where there is a public land register (although it is still complicated to accurately identify ownership completely), it is estimated that 50 percent of Scotland's privately held land is owned by just 432 individuals (Macfarlane 2021; Picken and Nicolson 2019). Identifying ownership is considerably more difficult in England. After considerable digging around, Guy Shrubsole has documented that around 550,000 acres (220,000 hectares) of England's peat moors are run for grouse shooting, and of these, around 300,000 acres (120,000 hectares) are comprised of just thirty estates. Of these, four are owned offshore (Shrubsole 2016).

The campaigns over grouse shooting, winter burning, and land access have routinely also been inflected with this politics of concentrated land ownership. Guy Shrubsole is the central campaigner in this regard,[3] and these campaigns have been popularized by journalists like George Monbiot (e.g. Monbiot 2020). Alongside emphasizing the high degree of land ownership inequality in the UK, exemplified by large privately owned moorland estates, these campaigns also demonstrate the character of the business model animating these estates. This business model is premised on elite consumption through annual grouse hunts. This has an economic dynamic in that it is the principal source of revenue generation for landowners (alongside the subsidies they get), although the economics of grouse shoots are difficult to clearly evaluate (RSPB n.d.), but also operates as an important process

[3] See the Who Owns England website (https://whoownsengland.org/) and particularly his report on the ownership of peat moors (Shrubsole 2016).

of elite political formation in the UK. Large landowners, often closely connected to the City of London financial sector,[4] use this potential to reproduce the elite prestige and power of landowners. Peat moors are basically exceptionally marginal lands in economic terms and have little other commercial potential than as grouse moors, and landowners' organizations present grouse shooting as the most sustainable possible economic use, suggesting that, if it is not possible, then landowners would put more damaging activities on the moors.[5]

Another key part of the business model for the moors is the land subsidy regime. The moors are counted and regulated as farmland. As such, while the UK was still in the European Union (EU) they received substantial subsidies under the EU Common Agricultural Policy, and continue to receive subsidies from other national schemes, and the regimes that are emerging post-Brexit. While there was some attempt within the UK government, most visibly by Michael Gove (at least publicly) and Zac Goldsmith (a junior minister in this context), to reorient such subsidies in the post-Brexit context toward environmental benefits rather than simple subsidies per acre of farmland owned, these attempts were largely abandoned during the political chaos of 2022, when the country was led by three different prime ministers. For the moment, landowners are heavily subsidized on a per acre basis. The moors are, of course, large areas, and landowners receive correspondingly large subsidies.

These questions of land ownership rarely figure in the accounts of peat restoration projects. In the *UK Peatland Strategy* document, for example (a forty-eight-page document), ownership is never mentioned, and only once is an individual landowner named, providing a testimonial for the benefits of peat restoration to them (IUCN 2018: 37). Occasionally it is mentioned in passing – the restoration project discussed in Vinter (2021), for example, is noted as "in partnership with a landowner." But the identity of the landowner is largely absent, and thus the nature of the tensions with private ownership, winter burning, and grouse shooting are unclear.

Conversely, however, land purchases for the purpose of carbon offsetting have had significant attention. Some particularly high-profile ones include BrewDog (the Edinburgh-based brewer of Punk IPA and other brands), which has bought 9,000 acres (3,600 hectares) in the Cairngorms to be able to claim negative carbon status (Carrell 2022a; Macfarlane 2021), involving both reforestation and peat restoration, as well as a distillery and eco-hotel complex (The Scotsman 2021). Similarly, large purchases by the Danish clothing billionaire Anders Holch Povlsen, now Scotland's largest individual landowner with at least 210,000 acres (84,000 hectares) of land,

[4] For example, the Moorland Association's president is currently Matt Ridley, former chair of the Northern Rock bank that collapsed in the financial crisis of 2007–2009, nephew of the former Conservative Secretary of State for the Environment Nicholas Ridley, owner of moorland in County Durham, and outspoken climate denier.

[5] For lowland peat, this is very different, and most such peat bogs have been long since drained, starting in the medieval period, and have become the UK's most productive agricultural land in East Anglia, Lincolnshire, and the like.

have entailed substantial reforestation and peatland restoration projects (Segal 2022). There is therefore a contradictory potential for concentrated land ownership to be deployed for peatland conservation, albeit with highly skewed social consequences (Carrell 2022a; Macfarlane 2021), generating more radical arguments for ending private landownership in peatlands (Heron and Heffron 2022; Ramsay 2022).

6.2.3 Leisure, Consumption, and the Cultural Politics of Peat

The final aspect of peat's politics that is brought into view by peat restoration initiatives is that peat moors in particular are a long-standing site of highly aestheticized leisure consumption (see Figure 6.1). These leisure activities include walking, climbing, fell running, and mountain biking, but also photography, painting, and nature documentaries. Accounts of these activities often refer back to a history of the cultivation of an aesthetic of upland Britain going back to the early nineteenth century Romantics, notably the poet William Wordsworth. These activities are now a very significant part of the rural economy in many parts of the UK, and significant parts of many people's leisure activities.

Alongside this general cultural economy of peatlands, it has also more specific cultural meanings and uses. A notable one of these is archaeology where "bog bodies" (Giles 2020) provide particularly powerful forms of archaeological evidence. Archaeologists also often now have embedded institutional authority to intervene and be consulted on development projects, as part of the regulation of heritage, and the impetus for peat or conservation is often stimulated by archaeological concerns (e.g. Brunning 2012). Another is the importance of peat to whisky production, where the prospect of "post-peat" whisky has been the object of both considerable cultural-economic anxiety given its centrality to Scottish identity in particular (Dick 2020), and considerable innovation in the industry (e.g. Fellows 2021), including the Bunnahabhain distillery, which was already marketing a peated whisky as "decarbonized" in time for COP27 in Glasgow (Carbon Ruins Scotland 2021).

The development of strong cultural attachments to and regular use of peat moors has been both a cause and a consequence of the political struggles over access to land for such walking. This politicization is, however, complicated by the relationship between the peat restoration initiatives and the politics of land access for walking. Since restoration involves rewetting large areas of moorland, it at least risks making them inaccessible for walking. At least in highly popular areas (Kinder Scout is again a classic case as it is so close to major urban areas like Manchester and Sheffield), there has been extensive path maintenance to prevent erosion since the 1970s, which would have to be continued and extended (and perhaps transformed from rock paths, which is the principal means now, to boardwalks across bogs), but which would of course transform the leisure experience itself of being

Figure 6.1 The author's dog, Fred, on Kinder Scout. Photo: Matthew Paterson. This picture is a good example of a dried-out and largely degraded peat moor environment, with heather as the dominant plant species. Successfully rewetted, it would become a blanket bog, retaining large amounts of water and storing carbon extremely effectively.

and walking on the moors, limiting their experience as "wild" spaces. There is consequently at least an implicit tension between the motivation of many to protect the peat moors and the actual consequences of that protection. Not all of the politicization of peat can be reduced to a class narrative of "the people" versus the landowners; it also has this more complex cultural political aspect, and is in effect also a conflict between competing forms of environmentalism (Macnaghten and Urry 1998) – an environmentalism focused on the aesthetic and leisure value of "wild spaces" versus a rather more technocratic environmentalism focused on the ecological value of peatlands due to their function in responding to climate change and improving biodiversity environmentalism.[6]

[6] There are more nuanced versions of this tension also. Among those involved in peat restoration projects, there is a visible tension generated by the way that these projects have been "carbonified" (Mert 2013; Methmann 2013), that is, framed increasingly around their climate change benefits to limiting carbon emissions or enhancing carbon uptake from the atmosphere. By contrast, the motivations of many involved in peat restoration are much more focused on the biodiversity benefits and the value of restoring ecosystems from their degraded state.

6.3 Conclusions

Peat restoration initiatives in the UK have made considerable progress. Many of the partnerships discussed in this chapter report quantities of carbon saved in terms of avoided emissions either from degraded peat and/or from carbon absorption as the peatlands are restored. But overall, the progress is so far relatively limited. Both at the UK level and specifically regarding Scotland, progress is largely inadequate. The CCC reported, for example, that Scotland was far from being on track to meet its admittedly very ambitious targets, and singled out peatland restoration as particularly weak, with only half of its annual restoration target of 20,000 hectares being achieved (Carrell 2022b; CCC 2022).

Explaining this weak level of achievement is to be sure in part because of inadequate funding, the inherent technical difficulties, and the like. But it is surely also because of the underlying political tensions over the ownership and use of peat moors that provide considerable constraints on the pursuit of a technocratic, depoliticized strategy embedded in the UK climate policy architecture, which is strongly oriented toward a stable set of policy instruments and infrastructure to manage a transition to net zero. This sort of tension is present in many areas of climate policy, as this volume as a whole attests, but the form it takes in relation to peat is in part because of the logic of *scope expansion*: the way that, as the stable policy regime in the UK has unfolded, new sources of emissions and carbon sinks have come into view, bringing novel sorts of power relation and political conflict with them. And given that the UK is ahead of most countries in terms of emissions reductions, other countries are likely to face similar problems as they expand the scope of climate policy beyond electricity supply.

References

Brulle, R. J. (2014). Institutionalizing delay: Foundation funding and the creation of US climate change counter-movement organizations. *Climatic Change* 122(4), 681–694.

Brunning, R. (2012). Partial solutions to partially understood problems: The experience of in situ monitoring and preservation in Somerset's Peatlands. *Conservation and Management of Archaeological Sites* 14(1–4), 397–405.

Bulkeley, H. (2016). *Accomplishing Climate Governance*. Cambridge: Cambridge University Press.

Bulkeley, H., Stripple, J., Nilsson, L. J. et al. (2022). *Decarbonizing Economies*. Cambridge: Cambridge University Press.

Carbon Ruins Scotland. (2021). Carbon ruins Scotland. Climaginaries. www.climaginaries.org/carbon-ruins-scotland

Carrell, S. (2022a). Lost forest: Why is BrewDog's green scheme causing controversy? *The* Guardian, March 5. www.theguardian.com/environment/2022/mar/05/lost-forest-why-is-brewdog-green-scheme-causing-controversy

Carrell, S. (2022b). Sturgeon told Scotland's climate targets are "in danger of being meaningless." *The* Guardian, December 7. www.theguardian.com/uk-news/2022/dec/07/sturgeon-told-scotlands-climate-targets-are-in-danger-of-being-meaningless

Carrington, D. (2016). Grouse shooting estates shored up by millions in subsidies. *The* Guardian, October 28. www.theguardian.com/environment/2016/oct/28/grouse-shooting-estates-shored-up-by-millions-in-subsidies

Carter, N. (2014). The politics of climate change in the UK. *WIREs Climate Change* 5(3), 423–433.

Carter, N., and Jacobs, M. (2014). Explaining radical policy change: The case of climate change and energy policy under the British Labour government 2006–10. *Public Administration* 92(1), 125–141.

Carter, N., and Little, C. (2021). Party competition on climate policy: The roles of interest groups, ideology and challenger parties in the UK and Ireland. *International Political Science Review* 42(1), 16–32.

Carter, N., and Pearson, M. (2024). From green crap to net zero: Conservative climate policy 2015–2022. *British Politics* 19(1), 154–174.

CCC (Climate Change Committee). (2022). *Progress in Reducing Emissions in Scotland: 2022 Report to Parliament*. Climate Change Committee. www.theccc.org.uk/wp-content/uploads/2022/12/Progress-in-reducing-emissions-in-Scotland-2022-Report-to-Parliament.pdf

CCC (Climate Change Committee). (2020). *Land Use: Policies for a Net Zero UK*. Climate Change Committee. www.theccc.org.uk/publication/land-use-policies-for-a-net-zero-uk/

Christophers, B. (2018). *The New Enclosure*. London: Verso.

Dick, S. (2020). The battle to save Scotland's peaty whisky in era of climate change. *Herald Scotland*, January 19. www.heraldscotland.com/news/18170966.battle-save-scotlands-peaty-whisky-era-climate-change/

Dubash, N. K., Pillai, A. V., Flachsland, C. et al. (2021). National climate institutions complement targets and policies. *Science* 374(6568), 690–693. https://doi.org/10.1126/science.abm1157

Evans, R. (2020). Grouse moors owners threatened government with legal action. *The Guardian*, February 20. www.theguardian.com/environment/2020/feb/20/grouse-moors-owners-threaten-government-with-legal-action

Ewert, S., and Abel, S. (2021). The transition to sustainable life on wetlands: How the sustainable use of Peatlands appears on the political agenda. In V. Beckmann, ed., *Transitioning to Sustainable Life on Land*. Basel: MDPI Books, pp. 175–193. https://doi.org/10.3390/books978-3-03897-379-4-8

Fankhauser, S., Averchenkova, A., and Finnegan, J. (2018). *10 Years of the UK Climate Change Act*. Policy report. London: Grantham Research Institute on Climate Change and the Environment, London School of Economics.

Farstad, F., Carter, N., and Burns, C. (2018). What does Brexit mean for the UK's climate change act? *Political Quarterly* 89(2), 291–297.

Farstad, F. M., Hermansen, E. A. T., Grasbekk, B. S., Brudevoll, K., and van Oort, B. (2022). Explaining radical policy change: Norwegian climate policy and the ban on cultivating peatlands. *Global Environmental Change* 74, 102517.

Fellows, R. (2021). Bruichladdich wants to turn your Whisky Green. *Esquire*, April 22. www.esquire.com/uk/food-drink/a36195013/bruichladdich-wants-to-turn-your-whisky-green/

Game and Wildlife Conservation Trust. (2014). Peat and heather moorland management. Position statement. www.gwct.org.uk/policy/position-statements/peat-and-heather-moorland-management/

Garavelli, D. (2022). Insight: Scotland's great Net Zero land grab. *The Scotsman*, April 17. www.scotsman.com/news/environment/insight-scotlands-great-net-zero-land-grab-3657133

Gewin, V. (2020). How peat could protect the planet. *Nature* 578(7794), 204–208.

Giles, M. (2020). *Bog Bodies*. Manchester: Manchester University Press.

Glasby, G. (2012). *Mass Trespass on Kinder Scout in 1932: And the Founding of Our National Parks*. Bloomington, IN: Xlibris.

Global Peatlands Initiative. (2021). Peatland Pavilion at COP26: IUCN UK Peatland Programme. IUCN Peatland Programme. www.iucn-uk-peatlandprogramme.org/events/peatland-pavilion-cop26

Goudarzi, S., Milledge, D. G., Holden, J. et al. (2021). Blanket peat restoration: Numerical study of the underlying processes delivering natural flood management benefits. *Water Resources Research* 57(4), e2020WR029209.

Hayes, N. (2021). *The Book of Trespass: Crossing the Lines That Divide Us*. London: Bloomsbury Publishing.

Heron, K., and Heffron, A. (2022). Renewing the land question: Against greengrabbing and green colonialism. *New Socialist*, February 20. http://newsocialist.org.uk/transmissions/renewing-land-question-against-greengrabbing-and-green-colonialism/

Hey, D. (2011). Kinder Scout and the legend of the Mass Trespass. *Agricultural History Review* 59(2), 199–216.

Hill, H. (1980). *Freedom to Roam: The Struggle for Access to Britain's Moors and Mountains*. Ashbourne: Moorland Publishing.

Horton, H. (2022a). Minister defends shelving right to roam report: "The countryside is a place of business." *The Guardian*, April 22. www.theguardian.com/environment/2022/apr/22/uk-minister-defends-shelving-of-right-to-roam-report-ahead-of-kinder-scout-trespass

Horton, H. (2022b). Burning ban failing to protect England's peatlands, say conservation groups. *The Guardian*, June 30. www.theguardian.com/environment/2022/jun/30/peatlands-in-england-at-risk-amid-apparent-flouting-of-regulations

IPCC (Intergovernmental Panel on Climate Change). (2014). Summary for policymakers. In *Climate Change 2014: Synthesis Report*. IPCC.

IUCN (International Union for Conservation of Nature). (2018). *UK Peatland Strategy*. IUCN. www.iucn-uk-peatlandprogramme.org/uk-strategy

Jackson, T. (2023). Rishi Sunak has ripped up decades of cross-party consensus on climate change. *The Conversation*, September 26. http://theconversation.com/rishi-sunak-has-ripped-up-decades-of-cross-party-consensus-on-climate-change-214287

Joosten, H., Sirin, A., Couwenberg, J., Jukka, L., and Smith, P. (2016). The role of peatlands in climate regulation. In A. Bonn, T. Allott, M. Evans, H. Joosten, and R. Stoneman, eds., *Peatland Restoration and Ecosystem Services: Science, Policy and Practice*. Cambridge: Cambridge University Press, pp. 63–76.

Kuzemko, C. (2016). Energy depoliticization in the UK: Destroying political capacity. *The British Journal of Politics and International Relations* 18(1), 107–124.

Lockwood, M. (2013). The political sustainability of climate policy: The case of the UK Climate Change Act. *Global Environmental Change* 23(5), 1339–1348.

Lockwood, M. (2021). A hard act to follow? The evolution and performance of UK climate governance. *Environmental Politics* 30(sup1), 26–48.

Macfarlane, L. (2021). Scotland is on the global frontlines of the Great Net-Zero Land Grab. *OpenDemocracy*, November 26. www.opendemocracy.net/en/oureconomy/scotland-is-on-the-global-frontlines-of-the-great-net-zero-land-grab/

Macnaghten, P., and Urry, J. (1998). *Contested Natures*. London: SAGE.

Mert, A. (2013). Discursive interplay and co-constitution: Carbonification of environmental discourses. In C. Methmann, D. Rothe, and B. Stephan, eds., *Interpretive Approaches to Global Climate Governance*. London: Routledge, pp. 23–39.

Methmann, C. P. (2011). The sky is the limit: Global warming as global governmentality. *European Journal of International Relations* 19(1), 69–91.

Ministry of Defence. (2022). MOD land holdings: 2000 to 2022. GOV.UK. www.gov.uk/government/statistics/mod-land-holdings-bulletin-2022/mod-land-holdings-2000-to-2022

Monbiot, G. (2020). If we want to cut flooding we should stop burning the moorland. *The Guardian*, February 11. www.theguardian.com/commentisfree/2020/feb/11/flooding-moorland-river-calder-flooded-peatlands

Natural England. (2021). Nature for Climate Peatland Grant Scheme. GOV.UK. www.gov.uk/guidance/nature-for-climate-peatland-grant-scheme

Newell, P., and Paterson, M. (1998). A climate for business: Global warming, the state and capital. *Review of International Political Economy* 5(4), 679–703.

Parker, G. (2008). Countryside access and the "right to roam" under New Labour: Nothing to CRoW about? In M. Woods, ed., *New Labour's Countryside: Rural Policy in Britain since 1997*. Bristol: Bristol University Press, pp. 135–148.

Paterson, M., Wilshire, S., and Tobin, P. (2023). The rise of anti-net zero populism in the UK: Comparing rhetorical strategies for climate policy dismantling. *Journal of Comparative Policy Analysis: Research and Practice* 26(3–4), 332–350. https://doi.org/10.1080/13876988.2023.2242799

Picken, A., and Nicolson, S. (2019). Who owns Scotland? The changing face of Scotland's landowners. *BBC News*, May 21. www.bbc.com/news/uk-scotland-47963208

Ramsay, A. (2022). Droughts and wildfires prove we need to end private land ownership. *OpenDemocracy*, August 16. www.opendemocracy.net/en/climate-crisis-land-ownership-wetlands-grouse-moor/

Ratamäki, O., Jokinen, P., Albrecht, E., and Belinskij, A. (2019). Framing the peat: The political ecology of Finnish mire policies and law. *Mires and Peat*, 24, 1–12.

Richards, P. (2021). Surge in peatland fires in UK national parks just weeks before climate summit. Greenpeace UK, October 12. www.greenpeace.org.uk/news/surge-in-peatland-fires-in-uk-national-parks-just-weeks-before-climate-summit/

Rowlatt, J. (2022). Peat soil fires: Campaigners say England's "rainforests" illegally burned. *BBC News*, May 29. www.bbc.com/news/science-environment-61607510

RSPB. (2022). Protecting England's peatlands: What this year's burning reports tell us. RSPB England, June 10. https://community.rspb.org.uk/ourwork/b/rspb-england/posts/protecting-england-s-peatlands-what-this-year-s-burning-reports-tell-us?CommentId=f40fd662-5f96-412b-a68e-b4098bfa96da

RSPB. (n.d.). Grouse shooting: Agriculture and land-use. RSPB England. www.rspb.org.uk/our-work/policy-insight/england-westminster/farming-and-land-use/driven-grouse-shooting/

Scottish Government. (2015). Scotland's national peatland plan: Working for our future. www.nature.scot/doc/scotlands-national-peatland-plan-working-our-future

Scottish Government. (2020). Securing a green recovery on a path to net zero: Climate change plan 2018–2032: Update. www.gov.scot/publications/securing-green-recovery-path-net-zero-update-climate-change-plan-20182032/

Segal, D. (2022). Who will profit from saving Scotland's Bogs? *New York Times*, May 5. www.nytimes.com/interactive/2022/05/05/headway/scotland-peatlands-climate-change.html

Shoard, M. (1987). *This Land Is Our Land: Struggle for Britain's Countryside*. London: Paladin.

Shrubsole, G. (2016). Who owns England's grouse moors? Who Owns England?, October 28. https://whoownsengland.org/2016/10/28/who-owns-englands-grouse-moors/

Shrubsole, G. (2019). *Who Owns England? How We Lost Our Land and How to Take It Back* (reprint ed.). London: William Collins.

Shuttleworth, E. L., Evans, M. G., Pilkington, M. et al. (2019). Restoration of blanket peat moorland delays stormflow from hillslopes and reduces peak discharge. *Journal of Hydrology X* 2, 100006.

Stokes, L. C. (2020). *Short Circuiting Policy: Interest Groups and the Battle over Clean Energy and Climate Policy in the American States*. Oxford: Oxford University Press.

The Scotsman. (2021). BrewDog steps up reforestation drive with purchase of Cairngorms estate. *The Scotsman*, April 15. www.scotsman.com/news/brewdog-steps-up-reforestation-drive-with-purchase-of-cairngorms-estate-3201687

UK Government. (2022). (n.d.). Sale of horticultural peat to be banned in move to protect England's precious peatlands. GOV.UK. www.gov.uk/government/news/sale-of-horticultural-peat-to-be-banned-in-move-to-protect-englands-precious-peatlands

Vinter, R. (2021). For peat's sake: The race is on to save Britain's disappearing moorland bogs. *The Observer*, September 12. www.theguardian.com/environment/2021/sep/12/for-peats-sake-the-race-is-on-to-save-britains-disappearing-moorland-bogs

7

Politicizing Financial Innovations for Transformative Climate Justice

JENNIE C. STEPHENS AND MARTIN SOKOL

The climate crisis is destabilizing the lives of most of humanity while also worsening inequities and disparities around the world. Within this context of more intense and frequent disruptions, the quest for societal stability and social justice requires transformative change in financial flows and financial structures. Yet the COVID-19 pandemic and early 2020s instability in global banking reveal how monetary policy and central bank actions further concentrate wealth and power among those who are already privileged. Monetary policies designed to promote stability in times of crisis have exacerbated economic inequities and disproportionally benefited billionaires and corporate interests while disadvantaging most people and sustaining the status quo (Sokol 2023). When governments around the world are willing and able to swoop in rapidly with policy changes to protect wealthy investors to ensure financial stability without considering how their policies are worsening climate instability and increasing economic precarity, there is a serious disconnect among stability in climate policies, stability in financial systems, and stability in the earth's climate systems (Chatterji and Stephens 2023).

Climate justice is an approach to climate action that goes beyond narrow efforts to reduce greenhouse gas emissions (Robinson 2018; Stephens 2022a). Climate justice centers the fact that climate change is inextricably connected to social justice, gender justice, racial justice, and economic justice because marginalized communities that have long faced underinvestment are more vulnerable to climate disruptions of all kinds, including extreme heat, flooding, storms, and food insecurity (Roberts-Gregory 2021). A commitment to climate justice is a paradigm shift from more conventional, technocratic ways of defining climate action (Sewerin

This chapter is partly based on our article in *Climate and Development* (Stephens and Sokol 2024) and our earlier conference paper (Sokol and Stephens 2022). We are grateful to the editors and contributors of this volume for their valuable and insightful comments and suggestions. Martin Sokol wishes to acknowledge that his research received funding from the European Research Council (ERC) Consolidator Grant under the European Union's Horizon 2020 research and innovation program (Grant Agreement No. 683197, project GEOFIN, of which he was a principal investigator).

et al. 2022), because climate justice focuses on investing in people and communities through initiatives that acknowledge and try to repair past environmental injustices while preventing continued and future climate injustices (Kashwan 2021; Stephens 2020). A commitment to climate justice is a commitment to societal transformation, because climate justice acknowledges that the climate crisis is a symptom of flawed economic and political systems that concentrate wealth and power through exploitative and extractive processes (Sultana 2022). A climate justice lens recognizes the reality that despite decades of dire warnings from climate scientists, political leaders around the world have been unable to resist the influence of corporate and financial interests and have inadvertently prioritized corporate interests over the public good (Stephens 2022b; Stoddard et al. 2021). An underappreciated point that we make in this chapter is that to achieve societal transformation for a stable, climate-just future, transformation of financial systems is required, which requires repoliticizing finance. This involves a new and different role for monetary policy, finance, and central banks, which have been considered apolitical in the large Western economies of the United States, the UK, and the European Union (EU), in line with the mantra of "central bank independence."

This chapter recognizes both synergies and tensions between the quest for stability of financial systems and the quest for stability in climate policies and highlights the need to (re)politicize central banks, monetary policy, and financial innovations. The chapter acknowledges antagonism in climate politics between two schools of thought. First, those advocating for a smooth transition guided by clear, predictable, and stable policies that can be leveraged by the wealthiest people and corporations. Second, those advocating for a more radical politics that disrupts and disempowers those who currently are the most privileged (Paterson et al. 2022), and for innovations in financial systems that could redirect society toward a more just, equitable, and stable future in the long term. The critical role of monetary policy, central banks, and managing financial risks in preparing for a more stable future is explained and explored, recognizing that this entails a repoliticization of monetary policy. Given that the people and institutions who would benefit most from financial innovations to advance climate justice do not have much power within current systems to facilitate and encourage transformative change, and recognizing that those who are profiting within the current systems are leveraging their power and influence to resist change and perpetuate the status quo, we propose a paradigm shift to reconceptualize stability and politicization in financial systems for climate justice. We explain how current mainstream depoliticized prioritization of financial stability is, in fact, worsening climate instability. We further argue that, conversely, current mainstream perspectives on climate policy stability are worsening long-term financial instability by reinforcing a depoliticized perspective on the role of financial institutions including central banks. We conclude by emphasizing that finance, central banks, and their monetary policies are an underappreciated part of climate politics and suggest that transformative climate policies require (re)politicization of finance and financial innovations.

7.1 Financialization and the (De)Politicization of Central Banks

Central banks – public institutions in charge of monetary policy (and often of banking regulation too) – are key to climate justice and are, therefore, a critical part of climate politics, yet they are under-analyzed (Sokol and Stephens 2022; Stephens and Sokol 2024). Only recently have some climate policy scholars begun to focus on monetary policy and how to link monetary policy with questions on how to respond to the chaos and instability of climate change and how to advance climate justice (Boneva et al. 2022; Kroll 2022; Langley and Morris 2020). The power of central banks to influence the financial systems has increased in the past four decades or so, in an era of financialization (Mader et al. 2020; Walter and Wansleben 2020), coinciding with the drive toward the depoliticization of monetary policy under the slogan of "central bank independence" (Fontan and Larue 2021). Financialization, a process of expanding the power of finance and debt in the economy (Lapavitsas 2013; Lazzarato 2012; Stockhammer 2008), has created a complex web of financial flows between creditors and debtors at multiple scales, and this ascendance of finance has provided a cover for depoliticization of wealth accumulation while also empowering banks and corporate interests to ignore the risks and inevitable instabilities of the climate crisis.

In response to a lack of theoretical tools to analyze the financialized, debt economy (Lazzarato 2012), the concept of "financial chains" was developed to elucidate the circulation of credit and debt in the economy (Sokol 2017, 2023; Sokol and Pataccini 2020). "Financial chains," defined both as channels of value transfer (between people and places) and as social relations of power (Sokol 2017), reveal the way in which financial architecture built on debt perpetuates inequality and instability. The "financial chains" framework highlights the central role of central banks in the financial system and in financialized capitalism more generally (Sokol 2023). Crucially, central banks set an interest rate and in doing so they manipulate the cost of debt throughout the economy. They lend money to banks and intervene in financial markets, for instance, by buying and selling government and corporate debt. In these ways, central banks are directly creating and managing "financial chains" with major implications for the rest of the economy. The importance of these interventions increases with increasing financialization of the economy. This has been highlighted by a series of crises, from the global financial crisis of 2008 to the COVID-19 pandemic of 2020 in which the monetary policy of central banks had major political implications. In fact, it would not be an exaggeration to say that without massive central bank interventions during these crises, the economic system would have probably collapsed (Sokol 2023). This central position of central banks in the economic system also makes them critical institutions for climate politics (Sokol and Stephens 2022). The critical role of central banks in these times of crises has led to greater political scrutiny of the politics of monetary policies.

7.2 Monetary Policy, Central Banks, and the Climate Crisis

Monetary policy refers to the decisions, actions, and communications of central banks as they influence the supply of money and the availability of credit in the economy (Boneva et al. 2022). The macroeconomic objectives of monetary policy tend to focus on price stability and financial stability; climate stability has not yet become an important objective of monetary policy (Dikau and Volz 2021). Only since the 2010s have some central banks begun to consider and integrate climate disruptions and the climate crisis into their policies and operations (Dafermos 2021; NGFS 2022; Volz 2017).

It is worth noting here that leading central banks of the Global North, including the United States, the UK, and the EU, are doing less to integrate climate disruptions and promote "green central banking" than their Global South counterparts (see, e.g., Barmes and Livingston 2021; Dikau and Ryan-Collins 2017; Volz 2017). In the Global South, monetary authorities are less constrained by the imperative of "central bank independence," which allows them to be more aligned with other national policy priorities, including responding to climate instability (Dikau and Ryan-Collins 2017). Some central banks in the Global South have thus introduced various measures to support the "green transformation" (Stephens and Sokol 2024). And the Chinese central bank has implemented a differential interest rate strategy for high/low-carbon investments (Kedward et al. 2022). In comparison, Western central banks are less advanced in "greening" their monetary policies, constrained by their price stability mandates and the goal of being "independent" from politics. The US Federal Reserve, the most powerful central bank in the world, is a case in point. In January 2023, the chair of the US Federal Reserve, Jerome H. Powell, declared that it would be "inappropriate for us to use our monetary policy or supervisory tools to promote a greener economy" (see Chatterji and Stephens 2023). This perception that responding to the climate crisis is too political for the central bank's monetary policy is, in fact, an ironic politicization of the stability mandate. By ignoring or dismissing climate destabilization in its policies, the US Federal Reserve is taking a political stance.

With rapidly accelerating climate instability and disruptions of all kinds, Langley and Morris have argued that central banks may well become "climate governors of last resort" (Langley and Morris 2020). Yet, despite some recent attention to the "greening" of the financial system, most central banks continue to support investment in climate-damaging fossil fuels (Dafermos 2021; Sokol and Stephens 2022). Financialized and seemingly "depoliticized" monetary policy also continues to concentrate wealth among corporate interests who then have more political power to resist policy action toward climate justice, so central banks are politicizing climate action while pretending to ignore the stabilizing need for climate policy. In multiple ways, central banks are in fact continuing to exacerbate climate instability and increase vulnerabilities to climate disruption (Sokol and Stephens 2022).

7.2.1 Debate on How Central Banks Should Respond

It is important to acknowledge that there is a growing debate about how central banks should respond to the climate crisis (e.g. Campiglio 2016; Campiglio et al. 2018; Corporate Europe Observatory 2016; Dafermos 2021; Gabor 2022; Monnin 2018; van 't Klooster 2021). Some argue that climate action is not part of central banks' mandate and that the responsibility for dealing with the climate crisis lies elsewhere (Skinner 2021). In the United States, climate change has become a divisive political issue, and the US Federal Reserve is usually seen as "apolitical" and "independent" of the rest of the government, so many argue that the Fed shouldn't get involved in political issues. In the meantime, and despite resistance from many corners, central bankers themselves are increasingly realizing that central banks can no longer avoid or ignore growing climate disruptions if they are to fulfill their primary objectives (Bolton et al. 2020; Carney et al. 2019; NGFS 2022; Svartzman et al. 2021). It is increasingly recognized that the climate crisis threatens the two main objectives of many central banks: monetary stability (the main element of which is price stability) and financial stability (including the resilience of the financial system as a whole) (Carney 2021: 90–91).

With regard to the *price stability* mandate (i.e. achieving low and stable inflation), many central banks seem to have not yet paid sufficient attention (although, on "fossilflation," see Schnabel 2022) to the volatility of energy systems reliant on unpredictable fossil fuels and food systems that are vulnerable to droughts, floods, and other climate disruptions (Chatterji 2022; Kuttner 2022). With increasingly complex geopolitics of fossil fuel supply, it is clear that the price volatility of fossil fuels is a critically important inflationary pressure (Kroll 2022; Melodia and Karlsson 2022), as witnessed by the current energy crisis. Given this, it would make sense for central banks to support the phaseout of fossil fuels in society (Chatterji 2022). Besides energy, food is another critical commodity and a major contributing factor to price instability. With worsening climate conditions for agricultural production, rising food prices will add to inflation, thus further highlighting the need for central banks to act on climate (Hertel 2016; Kuttner 2022).

The climate crisis is now increasingly challenging central banks' *financial stability* mandate too. As such, the threat to financial stability from climate disruptions is beginning to receive more attention from central bankers (Svartzman et al. 2021). Despite this growing attention, we are arguing that the financial stability mandate cannot be met unless and until long-term stability, including climate stability, replaces the current interpretation of this mandate, which has focused on short-term financial stability.

7.2.2 Green Swan Risks

Anticipated financial disruptions caused by the climate crisis are often referred to as "Green Swan" risks (Bolton et al. 2020; Svartzman et al. 2021). These financially disruptive events are projected to be the primary triggers of the next systemic financial crisis (Bingler and Colesanti Senni 2022) or worse. "Green Swans" (Bolton et al. 2020; Svartzman et al. 2021) are climate-induced "irreversible events triggering unpredictable chain reactions that are potentially catastrophic for the economy and financial system" (Svartzman et al. 2021: 564). Echoing the famous concept of the "Black Swan" by Taleb (2007) that foreboded the global financial crisis, "Green Swans" are considered "Climate Black Swans" with a potential to become much "more serious than most systemic financial crises" (Bolton et al. 2020; Svartzman et al. 2021).

A fundamental problem at the heart of many current central bank climate-related strategies in Western economies is that, for central banks, "it is the financial stability implications of climate change that to date have prompted their governmental interventions and proposals, and not the climate crisis itself" (Langley and Morris 2020: 1474). To put it crudely, under the current approach it does not seem to matter if planetary ecosystems are further destabilized, as long as systemically important financial institutions are able to hedge against the associated risks and the financial system as a whole stays more or less stable. The problem with this kind of approach is that eventually this will no longer be possible. Trying to stabilize a financialized system that is inherently unstable is problematic and is leading to all kinds of distortions of priorities. As Svartzman et al. (2021: 564, emphasis added) observe, most of the risk associated with "Green Swan" events "will remain *unhedgeable* unless a system-wide approach to the energy transition is undertaken."

Another major problem is that, despite recent attention to, and increasing rhetoric about, the "greening" of the financial system, most central banks continue to perform actions that undermine climate efforts (Sokol and Stephens 2022). In doing so, they are deepening the climate crisis and increasing the risks of more frequent and severe environmental, economic, and financial disruptions. Central banks continue to provide financial support to the fossil fuel industry, which allows continued fossil fuel exploration, extraction, and production. This has been most recently on display during the COVID-19 pandemic, which saw central banks (including the US Federal Reserve, the European Central Bank, and the Bank of England) supporting the fossil fuel industry both directly via unconditional quantitative easing (QE) and indirectly via its bank lending operations that lack any "green" conditionality. The direct channel (financial chain) involved the purchase of large quantities of corporate bonds by central banks as part of their QE. Following the "market neutrality" principle, these purchases simply reflected the current "market" and thus also included a large quantity of bonds of climate-damaging corporations. This effectively amounted to an unconditional direct subsidy for the fossil fuel industry. The so-called market neutrality, which lies at the heart of central bank operations, produces a strong bias toward fossil fuel energy that has been well documented (Boneva et al. 2022; Gabor 2022). The indirect channel involved central banks lending to commercial banks at extremely favorable rates (or even negative interest rates) without any conditions attached. In turn, commercial banks could lend this money (via cascading financial chains) as they see fit, including investing in projects that accelerate climate change.

Another fundamental issue is that central banks are narrowly interpreting their mandate and attempting to maintain (short-term) financial stability at all costs. In doing so, they stabilize financial markets and banking systems (and the attendant "financial chains"), thus perpetuating financialized systems that increase social inequality and deepen uneven development at various scales. This exacerbates climate vulnerabilities, so central banks are fostering conditions for future instability while compromising climate justice. Furthermore, major Western central banks focus on safeguarding financial stability in the Global North with little regard for the repercussions their actions will have on the Global South. Indeed, financial stability in the capitalist core can be achieved (temporarily) at the cost of economic, social, and environmental instability elsewhere, thus destabilizing the system globally. By focusing on maintaining short-term immediate financial stability, powerful central banks of the Global North are leading us toward more volatile instability in the long run (see also Stephens and Sokol 2024). Thus, from a climate justice lens, central banks are currently exacerbating human suffering around the world by stabilizing financialized economies in the short term while delaying the required transformation needed to achieve sustainability in the long term.

7.3 Central Banks at a Crossroads: Between Stability and (Re)Politicization

Central banks are at a crossroads. Echoing the tension between "stability" and "politicization" highlighted by Paterson, Tobin, and VanDeveer (Chapter 1, this volume), we consider the path forward for central banks. Clearly, there is growing recognition of the need for transformative change to move on a path toward climate justice, but whether the changes in financial systems should strive for stability or intentional disruption is an open debate (Paterson et al. 2022). Given that there are relatively few climate policy experts focusing on monetary policy and the role of central banks in climate policy (Bailey 2023; Boneva et al. 2022; Stephens and Sokol 2024), this is an underexplored and under-analyzed issue. Yet this is an issue that requires urgent attention.

For decades, central banks in leading Western economies have been operating according to the "central bank independence" mantra. As part of this, central banks have been considered depoliticized. Released from direct government control (and the political influence that goes with it), Western central banks have been operating as beacons of policy stability. For some, this policy stability is what successful climate policy should aspire to (Mildenberger and Lockwood, Chapter 13, this volume). However, in the case of central banks and their monetary policies, this policy stability largely works against climate objectives. Indeed, in the name of stability, central banks can be relied on to support financial markets and financial players at whatever cost. Policy stability pursued by central banks (plus a narrow interpretation of their mandates) means that central banks now seem unable to respond to the worsening climate instability that the world's economies are facing. This inability means that central banks are currently at odds with the need for a rapid and radical transformation. Repoliticizing central banks so that monetary policy is aligned with energy and climate policy would therefore seem necessary for the systemic financial changes that are needed for climate justice.

While some maintain that central banks should ignore climate issues and continue to use the same old tools in a futile attempt for short-term financial stability, others are increasingly recognizing how critical central banks are for transformative climate politics and climate policy. Among those who do recognize the central role of central banks, some may assume that simply aligning monetary policy with other climate policies or objectives can ensure a smooth and "stable" transition. However, given the scale and speed of transformation needed, this may be too little too late. Indeed, it is likely that a more radical approach may be required. One such approach could consist of an intentional, short-term "creative disruption" to the financial system led by central banks (Sokol and Stephens 2022). A short-term destabilization of the financial system may be necessary to make the financial reset for long-term stability and sustainability. Such a dramatic change requires repoliticizing monetary policy and broadening (and/or reinterpreting) the current "stability" mandates of central banks. New mandates would need to acknowledge that in a world of worsening climate chaos, long-term stability requires short-term disruption to steer humanity onto a different path toward a more stable, just, healthy, and sustainable future. To do this, central banks, monetary policy, and finance more broadly need to be (re)politicized. Acknowledging that an intentional repoliticization of central banks will have multiple consequences, it is important to recognize that central banks are already very political. With monetary policy consistently advantaging the most privileged in society while disadvantaging those who are already marginalized and vulnerable, the myth of "central bank independence" as is currently projected and perceived in most Western economies (Davies 2023) needs to be dismantled.

For a proposed "creative disruption" to have the desired effect of advancing climate justice, monetary policy would need to become part of an all-of-government approach to climate policy. This would mean that monetary policy would be (re) politicized, debated, and contested. Furthermore, implementing "creative disruption" for climate justice would require monetary policy to integrate climate action with social, economic, and environmental justice. In practical terms, this would require monetary policy to align with – and support – a range of other policies, including fiscal policy, energy policy, industrial and trade policy, social welfare policy, gender equality policy, and other social policies (Sokol and Stephens 2022).

This transformation in banking also means tackling head on the so-called climate paradox view (Carney 2021), which suggests that some choices may need to be made between addressing climate change or guaranteeing financial stability. Recent actions (from unconditional QE to raising interest rates which have benefited fossil fuel companies) demonstrate that when central banks are faced with the above dilemma, they have consistently chosen financial stability over climate stability. This preference for always prioritizing short-term financial stability accelerates climate chaos and thus contributes to the inevitability of much bigger financial instability ahead. It is time to resist and reconsider the traditional notion of financial stability and to reclaim and restructure the financial sector toward a more climate-just future. Rather than accepting the idea of a paradox or dilemma, it can be argued that long-term stability requires short-term disruption in the way monetary policy is implemented. Instead of reinforcing an artificial choice between climate stability and financial stability, there is a growing recognition that financial stability requires climate stability, and prioritizing long-term, environmental, social, and economic stability requires a paradigm shift in climate politics and climate policies.

A paradigm shift also requires a major change in how central banks are considering climate risks. Currently, there are two major categories of risk that central banks associate with climate change in their efforts to safeguard financial stability: physical risks and transition risks (Boneva et al. 2022, Kroll 2022; Langley and Morris 2020). Physical risks include financial losses of banks (and other financial institutions) from increasingly volatile climate-related disruptions and weather events (storms, floods, heat waves) as well as from impacts of long-term climate changes (e.g. sea level rise). Transition risks, on the other hand, include financial losses that financial institutions can suffer as a result of "a rapid low-carbon transition, including radical policy shifts, reputational impacts, technological breakthroughs or unexpected limitations, and shifts in market preferences and social norms" (Svartzman et al. 2021: 565). We argue that there are two major issues with the framing of these two categories of risk. First, this framing conceptualizes financial institutions as mere victims of the climate crisis, rather than recognizing that they are active participants in destabilizing climate (with the support of central banks, as discussed). Second, as long as "transition risks" include the apparent risks from policy shifts needed to decarbonize (and thus to reduce "physical risks"), there is a major contradiction at the heart of the efforts to achieve financial stability using these two types of risk as a guide. At the extreme, it is perfectly conceivable that powerful financial institutions and central banks can argue/lobby for slowing down climate action in the name of safeguarding (short-term, narrowly defined) financial stability.

Advocating an intentional disruption in financial systems goes against the prevailing current wisdom about how central banks should respond to the deepening climate crisis (which is focused on mitigating financial disruptions). It also challenges assumptions about "stability" and how stable and just futures can be achieved. However, given that transformative change is needed, it is likely that without some kind of politicized intentional disruption of how financial systems operate, climate instability and climate injustice will accelerate and worsen. Given the intersecting and cascading impacts of climate disruptions, a comprehensive agenda for large-scale transformation toward climate justice has to include a combination of policy instruments that result in coordinated investments in reducing climate vulnerabilities while simultaneously resisting fossil fuel extraction and reliance and supporting investment in a more equitable, healthy, and renewable-based future.

7.4 Conclusions

In this era of worsening climate crisis, all politics is climate politics (Aronoff et al. 2019). We argue, therefore, that the politics of our financial systems is climate politics and the monetary policies of central banks is also climate politics. For too long, central banks in the West have been hiding behind the mantra of "central bank independence" as supposedly apolitical institutions, without democratic scrutiny, decoupled from electoral cycles. As such, central banks have been ensuring "policy stability" of neoliberal economic management whose limits are now plainly visible. But this façade of depoliticized monetary policy is no longer viable or credible. Central banks and their monetary policies need to be repoliticized, and climate stability requires innovations in how financial systems are managed and structured.

Although climate politics has not yet adequately considered and integrated the critical role of finance and monetary policy, this is changing quickly as the monetary policy quest for "stability" is increasingly elusive, and the high cost of responding to climate disruptions and preparing for climate adaptation is being linked to finance (Kasdan et al. 2021; Kuhl et al. 2024). In this rapidly destabilizing world, climate politics needs to expand and reconceptualize the linkages among climate policy stability, financial stability, and climate justice.

We suggest that the effective repoliticization of finance, central banks, and their monetary policies requires a paradigm shift in how we conceptualize stability. Short-term stability that continues to concentrate wealth and power among a few is not in anyone's best interest in the long term. Given that the people and institutions who would benefit most from financial innovations to advance climate justice do not have much power within current systems to facilitate transformative change, repoliticization of monetary policy needs to focus on climate justice. Recognizing that those who are profiting within the current systems are leveraging their power and influence to resist change and perpetuate the status quo, political struggle and contestation are necessary.

As the climate crisis gets worse, the threats of multiple other destabilizing crises are simultaneously expanding. Pandemics, wars, inflation, species extinction, and resource scarcities are also destabilizing forces, which means we are living in a time of the "polycrisis" (Tooze 2022). When facing multiple interrelated crises, the simplistic, narrow priorities of the large Western central banks are completely insufficient in supporting long-term financial and societal stability. The limits of neoliberal monetary policies in dealing with the "polycrisis," let alone advancing climate justice, are now being fully exposed.

By emphasizing the critical role of monetary policy, finance, and central banks in the climate crisis, this chapter challenges mainstream assumptions regarding financial stability and depoliticized monetary policy and emphasizes how policy alignment and an all-of-government approach are necessary for transformation. When an all-of-government approach to transformative climate policies is embraced, new rules, different assumptions, and novel ways of thinking about financial stability and financial disruptions will emerge. A new political commitment to embracing the concept of "policy-mixes" is essential to link climate policies with monetary policy for the transformative societal changes that are needed for future societal stability.

References

Aronoff, K., Battistoni, A., Cohen, D. A. and Riofrancos, T. (2019). *A Planet to Win: Why We Need a Green New Deal*. London: Verso.

Bailey, D. (2023). "Building back better" or sustaining the unsustainable? The climate impacts of Bank of England QE in the COVID-19 pandemic. *British Politics* 19, 134–153.

Barmes, D. and Livingstone, Z. (2021). The green central banking scorecard: How green are G20 central banks and financial supervisors? Positive Money, March 31. https://positivemoney.org/publications/green-central-banking-scorecard/

Bingler, J. A. and Colesanti Senni, C. (2022). Taming the green swan: A criteria-based analysis to improve the understanding of climate-related financial risk assessment tools. *Climate Policy* 22(3), 356–370.

Bolton, P., Despres, M., Pereira da Silva, L. A., Samama F. and Svartzman, R. (2020). *The Green Swan: Central Banking and Financial Stability in the Age of Climate Change.* Basel: Bank for International Settlements.

Boneva, L., Ferrucci, G. and Mongelli, F. P. (2022). Climate change and central banks: What role for monetary policy? *Climate Policy* 22(6), 770–787.

Campiglio, E. (2016). Beyond carbon pricing: The role of banking and monetary policy in financing the transition to a low-carbon economy. *Ecological Economics* 121, 220–230.

Campiglio, E., Dafermos, Y., Monnin, P. et al. (2018). Climate change challenges for central banks and financial regulators. *Nature Climate Change* 8(6), 462–468.

Carney, M. (2021). *Value(s): Building a Better World for All.* Glasgow: Harper Collins.

Carney, M., Villeroy de Galhau, F. and Elderson, F. (2019). The financial sector must be at the heart of tackling climate change. *The Guardian*, April 17. www.theguardian.com/commentisfree/2019/apr/17/the-financial-sector-must-be-at-the-heart-of-tackling-climate-change

Chatterji, A. (2022). The Fed should not punish working people for inflation driven by big oil's greed. *Common Dreams*, August 1. www.commondreams.org/views/2022/2008/2001/fed-should-not-punish-working-people-inflation-driven-big-oils-greed

Chatterji, A. and Stephens, J. C. (2023). The Fed is out of touch on climate. *The Hill*, January 27. https://thehill.com/opinion/energy-environment/3833534-the-fed-is-out-of-touch-on-climate/

Corporate Europe Observatory (2016). The ECB's "quantitative easing" funds multinationals and climate change, December 12. https://corporateeurope.org/en/economy-finance/2016/12/ecb-quantitative-easing-funds-multinationals-and-climate-change

Dafermos, Y. (2021). How should central banks respond to the climate crisis? *SOAS Blog*, May 17. https://study.soas.ac.uk/how-should-central-banks-respond-to-the-climate-crisis/

Davies, H. (2023). The climate crisis threatens economic stability: Why are central bankers divided? *The Guardian*, January 20. www.theguardian.com/business/2023/jan/20/climate-economic-stability-central-bankers

Dikau, S. and Ryan-Collins, J. (2017). *Green Central Banking in Emerging Market and Developing Country Economies.* London: New Economics Foundation. https://neweconomics.org/2017/10/green-central-banking-emerging-market-developing-country-economies

Dikau, S. and Volz, U. (2021). Central bank mandates, sustainability objectives and the promotion of green finance. *Ecological Economics* 184. https://doi.org/10.1016/j.ecolecon.2021.107022

Fontan, C. and Larue, L. (2021). Central banking. In C. Borch and R. Wosnitzer, eds., *Routledge Handbook of Critical Finance Studies.* Abingdon: Routlegde, pp. 154–172.

Gabor, D. (2022). Green central banking. In Heinrich Böll Foundation, ZOE – Institute for Future-Fit Economies, and Finanzwende Recherche, eds., *Making the Great Turnaround Work: Economic Policy for a Green and Just Transition.* Berlin: Heinrich-Böll-Stiftung, pp. 39–45. https://eu.boell.org/en/making-the-great-turnaround-work-study

Hertel, T. W. (2016). Food security under climate change. *Nature Climate Change* 6(1), 10–13.

Kasdan, M., Kuhl, L. and Kurukulasuriya, P. (2021). The evolution of transformational change in multilateral funds dedicated to financing adaptation to climate change. *Climate and Development* 13(5), 427–442.

Kashwan, P. (2021). Climate justice in the global North: An introduction. *Case Studies in the Environment* 5(1). https://doi.org/10.1525/cse.2021.1125003

Kedward, K., Gabor, D., and Ryan-Collins, J. (2022). Aligning finance with the green transition: From a risk-based to an allocative green credit policy regime. UCL Institute for Innovation and Public Purpose, Working Paper Series (IIPP WP 2022-11). www.ucl.ac.uk/bartlett/public-purpose/publications/2022/jul/aligning-finance-green-transition

Kroll, M. (2022). Central banks, energy transition and controlling inflation. Policy note. World Future Council and LINGO.

Kuhl, L., Shinn, J. E, Arango-Quiroga, J., Ahmed, I. and Rahman, M. F. (2024). The liberal limits to transformation in the Green Climate Fund. *Climate and Development* 16(9), 826–837.

Kuttner, R. (2022). Climate costs are not inflation. *The American Prospect*, August 31.

Langley, P. and Morris, J. H. (2020). Central banks: Climate governors of last resort? *Environment and Planning A: Economy and Space* 52(8), 1471–1479.

Lapavitsas, C. (2013). *Profiting without Producing: How Finance Exploits Us All*. London: Verso.

Lazzarato, M. (2012). *The Making of the Indebted Man*. Cambridge, MA: MIT Press.

Mader, P., Mertens, D. and van der Zwan, N. (2020). *The Routledge International Handbook of Financialization*. London: Routledge.

Melodia, L. and Karlsson, K. (2022). Energy price stability: The peril of fossil fuels and the promise of renewables. Issue brief. All Economic Policy Is Climate Policy Series, Roosevelt Institute, May 11.

Monnin, P. (2018). Central banks and the transition to a low-carbon economy. Zurich, Council on Economic Policies, CEP Discussion Note 2018/1.

NGFS. (2022). Central Banks and Supervisors Network for Greening the Financial System. Website. www.ngfs.net/en.

Paterson, M., Tobin, P. and VanDeveer, S. D. (2022). Climate governance antagonisms: Policy stability and repoliticization. *Global Environmental Politics* 22(2), 1–11.

Roberts-Gregory, F. (2021). Climate Justice in the Wild n' Dirty South: An Autoethnographic Reflection on Ecowomanism as Engaged Scholar-Activist Praxis before and during COVID-19. In K. M. Q. Hall and G. Kirk, eds., *Mapping Gendered Ecologies: Engaging with and beyond Ecowomanism and Ecofeminism*. Lanham, MD: Lexington Books, pp. 125–147.

Robinson, M. (2018). *Climate Justice: Hope, Resilience, and the Fight for a Sustainable Future*. New York: Bloomsbury.

Schnabel, I. (2022). A new age of energy inflation: Climateflation, fossilflation and greenflation. Speech at a panel on "Monetary Policy and Climate Change" at The ECB and Its Watchers XXII Conference. Frankfurt, Germany March 17. www.ecb.europa.eu/press/key/date/2022/html/ecb.sp220317_2~dbb3582f0a.en.html

Sewerin, S., Cashore, B. and Howlett, M. (2022). New pathways to paradigm change in public policy: Combining insights from policy design, mix and feedback. *Policy & Politics* 50(3), 442–459.

Skinner, C. P. (2021). Central banks and climate change. *Vanderbilt Law Review* 74(5), 1301–1364.

Sokol, M. (2017). Financialization, financial chains and uneven geographical development: Towards a research agenda. *Research in International Business and Finance* 39, 678–685.

Sokol, M. (2023). Financialisation, central banks and "new" state capitalism: The case of the US Federal Reserve, the European Central Bank and the Bank of England. *Environment and Planning A: Economy and Space* 55(5), 1305–1324.

Sokol, M. and Pataccini, L. (2020). Winners and losers in coronavirus times: Financialization, financial chains and emerging economic geographies of the COVID-19 pandemic. *Tijdschrift voor Economische en Sociale Geografie* 111(3), 401–415.

Sokol, M. and Stephens, J. C. (2022). Monetary policy and ecological crisis: Towards a climate justice approach. Paper presented at 26th Forum for Macroeconomics and Macroeconomic Policy (FMM) Conference on "Post-Keynesian Economics and Global Challenges." Berlin, Germany, October 20–22.

Stephens, J. C. (2020). *Diversifying Power: Why We Need Antiracist, Feminist Leadership on Climate and Energy*. Washington, DC: Island Press.

Stephens, J. C. (2022a). Beyond climate isolationism: A necessary shift for climate justice. *Current Climate Change Reports* 8(4), 83–90.

Stephens, J. C. (2022b). Feminist, antiracist values for climate justice: Moving beyond climate isolationism. In J. Agyeman, T. Chung-Tiam-Fook and J. Engle, eds., *Sacred Civics: Building Seven Generation Cities*. New York: Routledge, pp. 177–189.

Stephens, J. C. and Sokol, M. (2024). Financial innovation for climate justice: Central banks and transformative climate disruption. *Climate and Development* 16(9), 762–773. https://doi.org/10.1080/17565529.2023.2268589

Stockhammer, E. (2008). Some stylized facts on the finance-dominated accumulation regime. *Competition & Change* 12, 184–202.

Stoddard, I., Anderson, K., Capstick, S. et al. (2021). Three decades of climate mitigation: Why haven't we bent the global emissions curve? *Annual Review of Environment and Resources* 46(1), 653–689.

Sultana, F. (2022). Critical climate justice. *The Geographical Journal* 188(1), 118–124.

Svartzman, R., Bolton, P., Despres, M. Pereira Da Silva, L. A. and Samama, F. (2021). Central banks, financial stability and policy coordination in the age of climate uncertainty: A three-layered analytical and operational framework. *Climate Policy* 21(4), 563–580.

Taleb, N. N. (2007). *The Black Swan*. New York: Penguin Random House.

Tooze, A. (2022). Welcome to the world of the polycrisis: Today disparate shocks interact so that the whole is worse than the sum of the parts. *Financial Times*, October 28. www.ft.com/content/498398e7-11b1-494b-9cd3-6d669dc3de33

van 't Klooster, J. (2021). *The ECB's Conundrum and 21st Century Monetary Policy: How European Monetary Policy Can Be Green, Social and Democratic*. Berlin: Transformative Responses to the Crisis, Finanzwende/Heinrich-Böll-Foundation.

Volz, U. (2017). On the role of central banks in enhancing green finance. Inquiry Working Paper No. 17/01. UNEP Inquiry into the Design of a Sustainable Financial System. Geneva, UNEP.

Walter, T. and Wansleben, L. (2020). How central bankers learned to love financialization: The Fed, the Bank, and the enlisting of unfettered markets in the conduct of monetary policy. *Socio-Economic Review* 18(3), 625–653.

8

Private Climate Governance and Policy Stability in the Financial Sector

VIRGINIA HAUFLER

> It is not light that we need, but fire; it is not the gentle shower, but thunder. We need the storm, the whirlwind, and the earthquake.
>
> —*Frederick Douglass, 1852*

Frederick Douglass powerfully criticized those who sought a more comfortable path to the abolition of slavery. In our current context, his words seem appropriate for addressing the climate crisis. The evidence that climate change has already affected our environment surrounds us, as millions of people suffer through heat waves, droughts, and floods. The rate of warming since 1981 has almost doubled, and the ten warmest years in recorded history have all occurred since 2010, with 2023 on track to be the hottest on record (National Oceanic and Atmospheric Agency 2023). A recent Intergovernmental Panel on Climate Change (IPCC) report noted the "brief and rapidly closing window of opportunity to secure a livable and sustainable future for all" (IPCC 2022). Activists including youth groups and scientists have taken to the streets and public square to demand action. To prevent further disaster will require large-scale transformation of the basis of our economy. This is not simply a technical problem but rather one of significant societal change.

One of the central debates among those pushing for action is whether they should adopt strategies that stabilize the policy environment, and if so, what the political impact of doing so would be. It seems paradoxical to seek change through stability – "stability" often denotes the status quo, which could mean either doing nothing about climate change or continuing current inadequate efforts. As Paterson,

This research is supported in part by a Fulbright Canada-Pierre Elliott Trudeau Foundation Fellowship. Fulbright Canada is a joint, bi-national, treaty-based organization to encourage mutual understanding between Canada and the USA through academic and cultural exchange. Fulbright Canada is supported by the Canadian Government through Global Affairs Canada, by the US Government through the Department of State, and by corporate sponsors, charitable trusts, and university partners. The Pierre Elliott Trudeau Foundation (PETF) is an independent nonpartisan charity founded as a living memorial to the former prime minister of Canada. The opinions and ideas expressed here are solely attributable to me and not to Fulbright Canada or the PETF.

Tobin, and VanDeveer discuss in their introductory chapter, stability can have other meanings: pathways to long-term emissions reductions, continuity in policy design, or engineering lock-in. Stable policy design can lock in practices and technologies that facilitate long-term emissions reductions pathways, linking together different goals. However, critics view this as a too-comfortable path that is inadequate for the challenge we face (Boykoff et al. 2010; Kouchakji 2023; Lamb et al. 2020).

In this chapter, I look at stability as policy lock-in with respect to the private sector, since industry is critical to the energy transition. Advocates argue that we must institutionalize norms and practices within industry that make it costly for business to maintain the current status quo. One prominent approach is through private governance initiatives, that is, voluntary industry self-regulation. These are collective voluntary standards-setting approaches that institutionalize principles, norms, and practices within a sector (Grabs et al. 2021; Vogel 2009). If industry collectively adopts common standards and practices, the argument is that this will prevent costly government regulation, create incentives for firms to comply, and help manage risks during the energy transition. Collective standards and practices can nudge firms off the old path and onto a new more sustainable one at lower cost and less pain.

Private governance initiatives narrow attention in ways that depoliticize debates by focusing on standards, reporting requirements, and appropriate technologies. The rise of global governance more generally, with its mixture of public and private authority, heralded the potential for a new kind of politics. Corporate social responsibility, private governance, and multistakeholder initiatives recast the role of business as partners in governance and not as political opponents. However, the promotion of industry-led climate initiatives is linked to political struggles within the climate movement, in which divisions over strategic choices can impact its effectiveness and coherence (Hadden 2015). These choices include the turn toward voluntary standards developed and adopted by the private sector. Green et al. (2021) argue that voluntary private governance reduces political conflict between NGOs and firms, but at the cost of slow – or even no – progress on climate mitigation or adaptation.

I argue here that private governance reflects a preference for policies that will provide a stable framework of expectations about the steps needed to make the energy transition and lock in particular standards and approaches. I explore this in the case of private climate governance in the financial sector. Banks, asset managers, and insurers all wield significant influence over policies and practices adopted by their customers. They set the conditions under which customers can borrow, invest, and manage risks. Therefore, the climate policies they adopt can ripple throughout the larger economy. For this chapter, I focus on the insurance industry and the Net Zero Insurance Alliance (NZIA), which is a private sector initiative among insurers to establish common commitments regarding decarbonization of their portfolio of business.

In what follows, I first discuss the concepts of stability and policy lock-in and their relationship to private governance. I then narrow my focus to the insurance industry, which is beginning to garner more attention for its climate policies. I provide a brief overview of how the insurance industry has responded to the challenge of climate change, its participation in the United Nations Environment Programme Finance Initiative (UNEP-FI), and the creation of the NZIA as a form of private governance. In the conclusion, I turn to the relationship between stability, private governance, and politicization.

8.1 Private Governance, Policy Stability, and Policy Lock-In

The idea of "lock-in," or path dependence, came out of the study of technological change, and the observation that initial technological choices constrain future options. Past decisions about how to design a car or a keyboard, for instance, make it costly to shift technologies as the context changes (David 1985; Lewin 2001; Liebowitz and Margolis 2012).[1] The same idea applies to policy adoption. For example, building highway infrastructure invites people to drive more and makes it harder to get support for policies that favor public transit. Policy lock-in figures in the literature on institutions and path dependence. The idea of *policy* lock-in has a narrower focus: policies are "locked in" when they are institutionalized, that is, when the initial choice of policy becomes a standard process that no longer requires a decision or choice. It is the default policy.

Unruh has written extensively on the idea of "carbon lock-in" as a form of negative path dependence (Seto et al. 2016; Unruh 2000, 2002). The policy choices about energy infrastructure taken in the past determine the technologies and energy sources we use today. This path dependence makes it extremely difficult to pursue an energy transition. Unruh argues that it is unlikely developing countries can develop further without fossil fuel-based industrialization because the carbon-based system has been so thoroughly globalized.

Climate activists often seek to create a *positive* path dependence by promoting policies that will do the opposite of carbon lock-in – policies that institutionalize alternative energy choices through support for new infrastructure, technology, and standards. Once in place, the goal is to move us irrevocably toward a green transition. In 2014, the IPCC report included a chapter on adaptation and implementation that highlighted institutionalization (Mimura et al. 2014). A year later, the Paris

[1] The concept of technological lock-in is a subject of contention in economics and law.

Agreement gave a greater role to sub-state and non-state actors to progressively reduce emissions, seeking to institutionalize their contributions (Hale 2016). In the succeeding years, institutions at all levels of government – international, national, and local – have developed climate adaptation and mitigation plans. These actions are embedding policies in institutions to create stable expectations regarding the direction of policy. As Paterson et al. (2022) note, the desire for this kind of policy stability can reduce political contention, with some observers labeling these solutions as "post-political." However, this approach sidesteps the need to do the political work to build coalitions supporting more significant change. Instead of pursuing policy stability, some argue we should be engaging in critical debates in the political arena. They ask whether we should "pump up the volume" instead of trying to avoid the messiness of mass politics (Adler and Kentikelenis 2022).

Private climate governance can be viewed as a way to institutionalize policy choices by firms. Grabs et al. (2021: 1183) define private governance systems as "the formulation of procedural and/ or substantive rules and standards by nongovernmental actors …, their monitoring and enforcement through the same actors or third parties, and the preferential treatment of actors in compliance with such rules, for example, through improved reputation, market access, pricing conditions, or access to financing." I include in private governance both the policies adopted by individual firms as they apply to their customers, clients, and business partners and the policies that are collectively negotiated and adopted.

There are four ways in which private climate governance is a strategy that favors stability and de-politicization:

1. *Private governance reinforces the market system.* Cashore, for instance, identifies a set of private initiatives that he calls "non-state market-driven" (NSMD) governance. These systems adopt processes such as certification and auditing to provide reputational benefits to participants. These benefits in turn provide an advantage in market competition, incentivizing participation and compliance (Auld 2014; Cashore et al. 2004; Grabs 2020; van der Ven 2019). Many voluntary initiatives rely on transparency as the mechanism for enforcement – firms are required to provide information revealing their compliance with standards set by the initiative. This approach assumes that public reporting will lead to a market response, such as consumers shunning a weakly performing company. Private governance thus relies on the market to lock in standards and stabilize policy choices.
2. *Private governance emphasizes consensus-building partnerships.* Pepermans and Maeseele (2016) note that climate politics multiplies antagonisms but also compels former opponents to partner. We see this in the collective nature of voluntary climate commitments among competing firms and in the proliferation

of public–private partnerships and multi-stakeholder initiatives (Westerwinter 2019). These are designed to combine the different competencies, resources, and authority of often antagonistic actors. These collective efforts institutionalize policies and reflect a strategy of stability. However, they are criticized for favoring superficial cooperation by adopting weak standards and emphasizing process over outcome (Gray and Purdy 2018; MSI Integrity 2020).

3. *Private governance empowers the private sector.* Industry has the resources, competence, and expertise required to achieve the energy transition. This capacity gives their participation in private climate governance a degree of legitimate authority (Haufler 2010). Firms have deep expertise regarding the specific ways in which their business generates greenhouse gas (GHG) emissions, and they have the technical capacity and organizational capability required for reducing them. When they make proposals for emissions reductions, they are viewed through this lens of expertise and capacity. In turn, this empowers firms in climate governance, which further reinforces market-based solutions, consensus-building partnerships, and technocratic problem-solving.
4. *Private governance favors process standards, technical indicators, and incrementalism.* Because they are based on voluntary membership, the initial standards in many private governance schemes must be low enough to attract members, although they also must be high enough to provide reputational benefits (Potoski and Prakash 2010). This is reinforced by the business penchant for benchmarking, that is, establishing standards to achieve over time. The standards are often about transparency such as reporting requirements and debates over indicators to report. This process rewards inching toward a desired goal over time instead of immediate achievement. Standards established at the creation stage may be raised over time as experience leads to acceptance and initial costs are absorbed, but slow steps forward protect the economic viability of the companies involved.

8.2 Insurance and Climate Change

The financial sector has long been viewed as a potential source of leverage to change the behavior of customers. Most attention focuses on the influence of investors and lenders, but the insurance sector has been a target of activists since the 1992 Rio conference (Paterson 2001). It is a critical element in adaptation to climate change, helping businesses and individuals assess risk, cushion losses, and strengthen resiliency. Insurers can signal a change in risk through shifts in pricing, deductibles, and insurance coverage (Surminski et al. 2019). Insurance itself favors stability because the financial cushioning it provides supports "business as usual" even in an era of disruption (Chandler and Coaffee 2016). Traditional risk models are backward-looking, based on historical data, and not oriented to future risks.

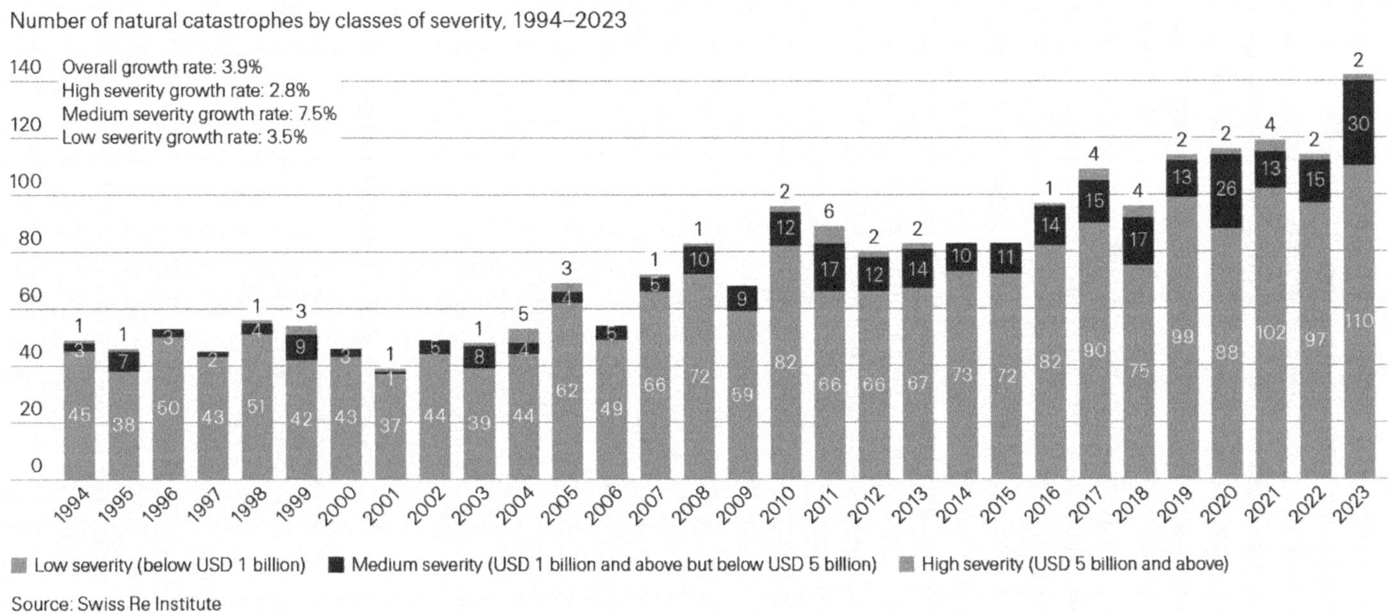

Figure 8.1 Number of natural catastrophes by classes of severity, 1994–23.

The industry is already experiencing costly payouts linked to a changing climate. The record for insured losses from natural disasters was set in 2017, reaching \$170 billion due to a particularly severe hurricane season in the United States. In 2021, insured losses from natural catastrophes added up to around \$130 billion globally, the fourth-highest on record (Reuters 2022).[2] The first half of 2023 experienced \$50 billion in losses, and Swiss Re – one of the world's largest reinsurance companies – estimates 68 percent were due to severe storms. Not only the number but the severity of natural disasters has increased (see Figure 8.1). Devastating floods and damage from high winds and excessive rainfall, extreme heat combined with severe drought, and massive wildfires around the world all may have been supercharged by global heating.

The insurance industry has two sources of leverage over other industries to promote climate mitigation and adaptation – through insurance underwriting contracts and through its investment portfolio ("Scope 3" emissions according to the Greenhouse Gas Protocol).[3] Insurers can set prices (premiums) and conditions for coverage in their insurance business and selectively invest income from insurance premiums in their investment portfolios. They are on both sides of the asset revaluation dynamic laid out by Colgan et al. (2021): They are climate-forcing asset holders on the investment side because they can impose conditions on the firms in which they invest. They are climate-vulnerable asset holders as insurers because they insure assets that are vulnerable to climate impacts. On the investment side, their choice of where to invest can influence who gets financing – fossil fuel projects versus renewables. On the insurance side, their choices about how to evaluate and price risk or what behavioral conditions to include in contracts influence customer choices about home construction or transportation. The most powerful signal they can send about how they understand risk is by withdrawing insurance coverage entirely from activities or locations at high risk.

8.3 Private Climate Governance, Insurance, and Policy Stability

Private governance in insurance is process-oriented, focused on technical standards, and relies on market incentives to encourage compliance. Meeting standards typically means measuring and reporting emissions. Ideally it requires firms to revise insurance products and policies to incentivize climate-sensitive practices by clients and set prices (premiums) to reflect forward-looking climate risk. If

[2] These figures include earthquake disasters. Notably, estimates by Munich Re are that 40 percent of natural disasters worldwide are not covered by insurance.

[3] The Greenhouse Gas Protocol defines Scope 3 emissions as "the result of activities from assets not owned or controlled by the reporting organization, but that the organization indirectly impacts in its value chain." (US Environmental Protection Agency 2023). The Protocol itself is a set of standards for measuring emissions developed by the World Resources Institute and the World Business Council for Sustainable Development.

these policies become institutionalized as standard policy clauses, they will lock in climate mitigation and adaptation by customers. For instance, customers may be offered better prices and coverage if they strengthen roofs and waterproof basements in storm-prone locations. This approach focuses on modifications to reduce losses and technocratic models of risk. It is not intended to redistribute costs and benefits but rather emphasizes business as usual. This result is policy stability and continuity with existing practices (Paterson 2021).

Many insurers participate in private climate governance and public–private partnerships addressing climate issues. The overarching framework for their commitments is the United Nations Environment Programme Finance Initiative (UNEP-FI), founded in 1992. UNEP-FI established standards on sustainable finance for bankers, investors, and insurers. But insurers belong to other initiatives: the UN Global Compact, the Principles for Responsible Investment (PRI), the Equator Principles, the Poseidon Principles for Marine Insurance, and the UN Sustainable Blue Finance Initiative. The most relevant one for examining the relationship between private climate governance and policy stability is the Net Zero Insurance Alliance (NZIA), which was orchestrated by the UNEP-FI.

In 2021, criticism of climate action by the financial sector centered in part on how financial firms were all pursuing different climate policies, targets, and indicators. Leading CEOs formed a new partnership, the Glasgow Financial Alliance for Net Zero (GFANZ), to bring different initiatives under one umbrella group. Members would align their efforts with the UN Race to Zero campaign, which means their policies are science-based, cover all emissions, have interim targets, and commit to transparency. Under this umbrella, different sectors – banks, asset managers, and insurers – launched collective commitments to net zero.

Facilitated by the GFANZ, the NZIA was launched by some of the largest international insurers in July 2021. These included Allianz, a German company with more than 150,000 employees earning around 754 billion euros in premiums in 2020; AXA, a French firm that is even bigger; Lloyd's, a private company based in the UK; the big reinsurers Munich Re and Swiss Re; and firms based in Italy, Spain, Kenya, South Korea, and Japan (Cox 2022a). The NZIA commits members to achieve net zero emissions in their insurance and reinsurance portfolios *by their customers and business partners* by 2050, with interim targets every five years after 2030. In other words, the commitment is not about their own carbon emissions but those they insure. Insurers are required to report regularly on their progress. All members must also sign the UNEP Principles for Sustainable Insurance, align with the 1.5°C ceiling established by the Paris Agreement, and advocate for science-based and socially just transition policies. Participants would be delisted if they did not meet deadlines (Cox 2022a, 2022b).

Initially, they were stymied by a lack of consensus on how to measure the emissions of their customers (Scope 3 emissions). They could not set net zero targets for underwriting because they had no common standard for measuring the emissions of those they insure. This led NZIA to work with the Partnership for Carbon Accounting Financials to develop metrics, which were published in November 2022. Insurers will use them in developing insurance contracts and pricing policies going forward. They will have to develop new underwriting standards to include these metrics, particularly for the GHG-intensive activities of clients. They also plan to develop incentives for reducing emissions, such as special insurance for new technologies, nature-based solutions, and claims management (Cox 2022a, 2022b; UNEP 2021).

One goal of the NZIA was to change how insurers evaluate climate risks in their underwriting. Insurance models are backward-looking, with risk estimates based on historical experience with losses. If natural disasters in the past were few and far between, then the evaluation of risk in long-term insurance contracts assumes the probabilities remain the same. This is one element of the conservative character of the industry, binding it to historical experience. The industry now recognizes the need to develop future-oriented models that incorporate climate change.

We can identify some of the characteristics that I argue illustrate the ways in which strategies of private climate governance are oriented toward policy stability, locking in particular approaches to action. First, insurance itself is a market-based lever to induce customers to reduce GHG emissions. The NZIA was launched in part due to the recognition by executives of the market signals about climate change, such as recent insured losses. Policymakers such as US Special Envoy for Climate John Kerry lauded the GFANZ as evidence that financial firms recognized the commercial opportunities of the energy transition (Jessop 2021). These all indicate that the "business case" for pursuing net zero was a significant incentive for participants. In turn, the NZIA adopted practices that would preserve the market and not overturn it.

Second, the NZIA, and the GFANZ which facilitated it, is a partnership among its corporate members. It is embedded within the UNEP-FI and linked to a number of other private governance efforts, reflecting a preference for public–private consensus building (Dubash 2021; Mills 2005; Mills and Lecomte 2006). NZIA members must commit to the UNEP Principles on Sustainable Insurance, launched in 2012. Its leaders have expressed public support for the recommendations of the Task Force on Climate-related Financial Disclosures led by Michael Bloomberg, and for emerging frameworks such as the Task Force on Nature-related Financial Disclosures, in addition to the United Nations Sustainable Development Goals (UNSDGs) and the Global Biodiversity Framework. This complex network favors consensus-building partnerships that obscure some of the fundamental conflicts within the private sector.

Third, the NZIA empowers the private sector by leaving it to the insurers to develop standards and set their own goals. Each firm develops their own approach to reducing GHG emissions to avoid anti-trust accusations. The NZIA says nothing about divesting from fossil fuel projects entirely, which would do the most to achieve net zero commitments. The idea of "net zero" itself reflects the ability of firms to redirect attention away from hard goals. Net zero allows industry to continue with GHG emissions in some activities by offsetting them with reductions elsewhere. The NZIA lets insurers themselves off the hook, as it does not include provisions about GHG emissions by the insurers themselves.

Fourth, the NZIA focused a lot of its attention on reporting standards, emphasizing the process involved in commitments to transparency. Technical debates over how to evaluate the emissions of clients were an early stumbling block. Unlike other initiatives, the NZIA has established interim targets between now and 2050, but overall, it favors gradual approaches. The timeline to achieve net zero is slow and aspirational and involves regular reporting and incremental improvements. According to one industry representative, insurance is too often viewed only as a way to absorb financial shocks, leaning more toward adaptation and resilience than mitigation and decarbonization (Cox 2022a, 2022b).

8.4 Disruptions and Repoliticization?

The NZIA does not *require* participants to stop insuring fossil fuel projects – but strikingly, many are doing so, as they recognize the rising risks and costs of these projects (Insure Our Future 2022). Their decisions have been reinforced by clear signals from governments, particularly in Europe. The founding members of the NZIA include firms that started to phase out coal from their insurance portfolio some years ago. Swiss Re and Zurich Insurance led the charge to end coal insurance and also began divesting from coal projects. Lloyd's of London has committed to end insurance for coal-fired power plants, although not all Lloyd's members are going along with this. AIG adopted a comprehensive new coal and tar sands exit policy, committed to ending insurance for new projects and customers, and will phase out existing customers by 2030. Munich Re recently committed to end insurance for new fossil fuel projects and established new climate targets for its clients. Around thirty-eight major insurers are exiting coal (Insure Our Future 2022). Some observers view the end of insurance for coal projects as a sign that it will be phased out entirely as a major energy source. However, this optimistic take is undermined by big players such as Berkshire Hathaway that continue to cover these projects, and by the recent spurt of coal development projects in China (Centre for Research on Energy and Clean Air 2023).

Insurers are also withdrawing insurance from areas affected by extreme natural disasters, most notably in the United States. In Florida, property insurance has become harder to find as repeated storms and flooding have caused significant losses to insurers. Small firms have gone out of business while the major players have withdrawn. The same is happening in California due to wildfire risks and is spreading to Texas and other states. Insurance rates have gradually increased in response to a recalculation of future risk. Recent reports express concern that the insurance industry itself is being destabilized by climate change, and this could have wider implications for financial stability in general (Frank 2023). The insurance industry is shifting its approach to one that takes climate risk into account, instead of looking at the past for data on risk.

The combination of losses from natural disasters linked to climate change, and efforts to develop common policies and approaches to address it, has led to two kinds of backlash. Within the climate advocacy community, a more prominent radical wing has mobilized against the weakness of efforts by insurers. For instance, net zero commitments are viewed as particularly weak, and are unlikely to be achieved in time to prevent irrevocable changes (Dyke et al. 2021; Wilkes 2023; Wilkes et al. 2023). While groups such as Extinction Rebellion get a lot of media attention for their tactics, groups targeting insurers have become more active. The NGO Insure Our Future campaigns to get insurers to end completely all insurance for coal, oil, and gas. It scores companies on their efforts – or lack of them – and engages in public protests outside insurance company headquarters. This is a shift from previous strategies.

But the more significant political contention comes from outside the climate advocacy community and from insurers themselves. A recent article in an industry journal asked in its title, "Are insurers being bullied?" (Moorcraft 2022). While insurers may view climate advocates as too pushy, it is the climate deniers who pose the greatest threat. Insurers have become targets of the "anti-ESG," anti-"woke" culture wars pursued by the radical right in some US states ("ESG" is shorthand for environmental, social, and governance indicators that are often used as a means of labeling "ethical" investment portfolios). Some states in the United States now ban government agencies from working with financial firms that incorporate ESG standards into their decision-making or who offer ESG products. Texas recently banned insurance companies from using ESG criteria in setting premiums, thus excluding climate risk from consideration (Ahmed 2023; Malone et al. 2023). Others such as North and South Dakota are considering similar legislation targeting insurers, while around twenty states have passed broad anti-ESG bans. In contrast, California is considering legislation that would require incorporating ESG considerations into business decisions (Wolman and kahn 2023).

These local political and legal threats have had global impact. State attorneys-general in twenty-three US states warned members of the NZIA that their collaboration could violate US anti-trust laws, threatening to pursue litigation. In response, member firms left the NZIA, including some of its founding companies. In July 2023, the NZIA chose to end the requirement for members to set GHG reduction targets in an effort to avoid the political contention, but the group was down to only fourteen members at that point (Fellowes-Granda 2023). The GFANZ, the global umbrella for eight net zero industry groups, also has been undermined by these anti-trust threats and corporate withdrawals.

The insurance industry, and the financial sector as a whole, have been buffeted from both sides during the 2020s. Their lagging and weak response to demands for action on climate change, despite increased losses due to natural disasters, along with the potential collapse of their most visible effort – the NZIA – could lead climate advocates to repoliticize the issues. At the same time, even those weak efforts have been too much for the deeply politicized divisions fomented by climate deniers. The political response to the stability embedded within private governance systems may take the shape of a U-curve. The NZIA demonstrates this dynamic, at least with respect to the United States. We may see cycles of stability and repoliticization in the relations between industry and activists, a dynamic that Paterson, Tobin, and VanDeveer highlight in their introductory chapter.

The policies of the financial sector can be a powerful force for policy stability and lock-in. At the same time, they can stimulate a powerful political backlash, especially when they are reinforced by the costly impact of a warmer climate. While private industry pursues incremental change, many people today would argue that we need contentious political debate to generate the political will to take the costly steps that are needed to prevent the climate crisis from getting worse. At the start of this chapter, I posted a quotation from Frederick Douglass, the noted Black abolitionist and statesman. He framed the issue of slavery with a powerful analogy that resonates when it comes to climate action. We do not need the light, we need the fire; not the gentle rain, but the thunder.

References

Adler, D. R. K. and Kentikelenis, A. E. (2022). Pump up the volume: From covert to overt politics in global governance. *Global Perspectives* 3(1), 1–11.

Ahmed, A. (2023). Lawmakers pass bill to stop insurers from considering ESG criteria in setting rates. *The Texas Tribune*, June 12.

Auld, G. (2014). *Constructing Private Governance: The Rise and Evolution of Forest, Coffee, and Fisheries Certification*. Yale Agrarian Studies Series. New Haven, CT: Yale University Press.

Boykoff, M. T., Frame, D. and Randalls, S. (2010). Discursive stability meets climate instability: A critical exploration of the concept of "climate stabilization" in contemporary climate policy. *Global Environmental Change* 20(1), 53–64.

Cashore, B., Auld, G. and Newsom, D. (2004). *Governing through Markets: Forest Certification and the Emergence of Non-state Authority*. New Haven, CT: Yale University Press.

Centre for Research on Energy and Clean Air. (2023). China permits two new coal power plants per week in 2022. Briefing, February. Centre for Research on Energy and Clean Air.

Chandler, D. and Coaffee, J., eds. (2016) *The Routledge Handbook of International Resilience*. London: Routledge.

Colgan, J. D., Green, J. F. and Hale, T. N. (2021). Asset revaluation and the existential politics of climate change. *International Organization* 75(2), 586–610.

Cox, T. (2022a). Insurers begin long countdown to net zero. Environmental Finance, March 4.

Cox, T. (2022b). Insurers begin Long Countdown to Net Zero: Part Two. Environmental Finance, March 7.

David, P. A. (1985). Clio and the economics of QWERTY. *The American Economic Review* 75(2), 332–337.

Dubash, N. (2021). Varieties of climate governance: The emergence and functioning of climate institutions. *Environmental Politics* 30(suppl), 1–25.

Dyke, J., Watson, R. and Knorr, W. (2021). Climate scientists: Concept of net zero is a dangerous trap. *The Conversation*, April 22.

Fellowes-Granda, L. (2023). Insurers' failure on climate disasters imperils financial system. *The Washington Post*, July 11.

Frank, T. (2023). Climate change is destabilizing the insurance industry. *Scientific American*, March 23.

Grabs, J. (2020). *Selling Sustainability Short? The Private Governance of Labor and the Environment in the Coffee Sector*. Cambridge: Cambridge University Press.

Grabs, J., Auld, G. and Cashore, B. (2021). Private regulation, public policy, and the perils of adverse ontological selection. *Regulation & Governance* 15(4), 1183–1208.

Gray, B. and Purdy, J. (2018). *Collaborating for Our Future: Multistakeholder Partnerships for Solving Complex Problems*. Oxford: Oxford University Press.

Green, J., Hadden, J., Hale, T., and Mahdavi, P. (2021). Transition, hedge, or resist? Understanding political and economic behavior toward decarbonization in the oil and gas industry. *Review of International Political Economy*, 1–28. https://doi.org/10.1080/09692290.2021.1946708

Hadden, J. (2015). *Networks in Contention: The Divisive Politics of Climate Change*. Cambridge Studies in Contentious Politics. Cambridge: Cambridge University Press.

Hale, T. (2016). "All hands on deck": The Paris Agreement and nonstate climate action. *Global Environmental Politics* 16(3), 12–22

Haufler, V. (2010). Corporations in zones of conflict: Issues, actors, and institutions. In D. Avant, M. Finnemore, and S. Sell, eds., *Who Governs the Globe?* Cambridge: Cambridge University Press, pp. 102–130.

Insure Our Future. (2022). With new coal uninsurable, insurers start to move on oil and gas. Press release, October 19. https://global.insure-our-future.com/with-new-coal-uninsurable-insurers-start-to-move-on-oil-and-gas/

IPCC (Intergovernmental Panel on Climate Change). (2022). *Climate Change 2022: Impacts, Adaptation, and Vulnerability*. Working Group II Contribution to the Sixth Assessment Report of the Intergovernmental Panel on Climate Change. Cambridge: Cambridge University Press.

Jessop, S. (2021). Carney, Kerry launch global finance plan to boost climate action. *Reuters*, April 21.

Kouchakji, K. (2023). The climate is changing faster than we are. Energy Monitor, July 19. www.energymonitor.ai/opinion/opinion-climate-action-progress-too-slow/?cf-view&cf-closed

Lamb, W. F, Mattioli, G., Levi, S. et al. (2020). Discourses of climate delay. *Global Sustainability* 3(e17), 1–5.

Lewin, P. (2001). The market process and the economics of QWERTY: Two views. *The Review of Austrian Economics* 14(1), 65–96.

Liebowitz, S. J. and Margolis, S. E. (2012). The troubled path of the lock-in movement. *Journal of Competition Law & Economics* 9(1), 125–152.

Malone, L., Holland, E. and Houston, C. (2023). ESG battlegrounds: How the states are shaping the regulatory landscape in the US. Harvard Law School Forum on Corporate Governance, March 11.

Mills, E. (2005). Insurance in a climate of change. *Science* 309, 1040–1044.
Mills, E. and Lecomte, E. (2006). *From Risk to Opportunity: How Insurers Can Proactively and Profitably Manage Climate Change, August 2006*. Boston, MA: CERES.
Mimura, N., Pulwarty, R. S., Duc, D. M. et al., eds., (2014). Adaptation planning and implementation. In *Climate Change 2014: Impacts, Adaptation, and Vulnerability*. Part A: Global and Sectoral Aspects. Working Group II Contribution to the Fifth Assessment Report of the Intergovernmental Panel on Climate Change. Cambridge: Cambridge University Press.
Moorcraft, B. (2022). Drop fossil fuel clients: Are insurers being bullied? Insurance Business America, February 22.
MSI Integrity. (2020). *Not Fit-for-Purpose: The Grand Experiment of Multi-stakeholder Initiatives in Corporate Accountability, Human Rights and Global Governance*. MSI Integrity report. www.msi-integrity.org/not-fit-for-purpose/
National Oceanic and Atmospheric Agency. (2023). Climate change: Global temperature. Climate.gov News, January 18. www.climate.gov/news-features/understanding-climate/climate-change-global-temperature
Paterson, M. (2001). Risky business: Insurance companies in global warming politics. *Global Environmental Politics* 1(4), 18–42.
Paterson, M. (2021). Climate change and international political economy: Between collapse and transformation. *Review of International Political Economy* 28(2), 394–405.
Paterson, M., Tobin, P. and VanDeveer, S. D. (2022). Climate governance antagonisms: policy stability and repoliticization. *Global Environmental Politics* 22(2), 1–11.
Pepermans, Y., and Maeseele, P. (2016). The politicization of climate change: Problem or solution? *WIREs Climate Change* 7(4), 478–485. https://doi.org/10.1002/wcc.405
Potoski, M., and Prakash, A., eds. (2010). *Voluntary Programs: A Club Theory Perspective*. Cambridge, MA: MIT Press.
Reuters. (2022). Climate change "steroid" contributed to $130 billion in insured losses last year-Aon. *Reuters*, January 25.
Seto, K. C., Davis, S. J., Mitchell, R. B. et al. (2016). Carbon lock-in: Types, causes, and policy implications. *Annual Review of Environment and Resources* 41(1), 425–452.
Surminski, S., Barnes, J. and Vincent, K. (2019). Insurance as a catalyst for government climate planning? A framework for analyzing drivers and barriers, tested against evidence emerging from Sub-Saharan Africa. Working Paper No. 359. Centre for Climate Change Economics and Policy, University of Leeds and London School of Economics and Political Science.
UNEP. (2021). Global insurance and reinsurance leaders establish groundbreaking net-zero alliance. UNEP, July 11.
Unruh, G. C. (2000). Understanding carbon lock-in. *Energy Policy* 28(12), 817–830.
Unruh, G. C. (2002). Escaping carbon lock-in. *Energy Policy* 30(4), 317–325.
US Environmental Protection Agency. (2023). *Scope 3 Inventory Guidance*. www.epa.gov/climateleadership/scope-3-inventory-guidance
van der Ven, H. (2019). *Beyond Greenwash? Explaining Credibility in Transnational Eco-Labeling*. Oxford: Oxford University Press.
Vogel, D. (2009). The private regulation of global corporate conduct. In W. Mattli and N. Woods, eds., *The Politics of Global Regulation*. Princeton: Princeton University Press, pp. 151–188.
Westerwinter, O. (2019). Transnational public-private governance initiatives in world politics: Introducing a new dataset. *Review of International Organizations* 16, 1–38.
Wilkes, T. (2023). UN climate alliance scraps rules for insurers after exodus. *Reuters*, July 6.
Wilkes, T., Hübner, A. and Sims, T. (2023). Insurers flee climate alliance after ESG backlash in the US. *Reuters*, May 26.
Wolman, J. and Kahn, D. (2023). Anti-ESG comes for the insurance industry. *Politico*, March 7.

Part III

Comparative Politics

9

Energy Transition in Brazil and South Africa

Policy Stability vs. Politicization

KATHRYN HOCHSTETLER

Does meaningful decarbonization depend on policy stability that makes climate policies and institutional development irreversible, or does it depend on mastering a messy political conflict with steps forward and backward that might be inherent in large-scale political economy transitions? This chapter uses the example of energy transition in Brazil to suggest that policy stability in a bureaucratic and political status quo can in fact result in significant change, exemplified in the dramatic growth of wind power there since 2002. However, the Brazilian example also shows that the results were limited. Notably, solar power languished for a decade after wind power took off. Further, oil and gas production in Brazil expanded even faster than the new renewables. Thus, while a status quo policy stability enacted through bureaucratic politics led to some important changes in the Brazilian energy system, it proved unable to galvanize a full energy transition.

The delay to a full transition does not unequivocally point to politicization as a clear strategy either, however. A brief contrasting case study of South Africa, where politicization dominated the storyline as the country tried to move from its historic dependence on coal, shows the difficulty transformative forces faced as they took on powerful incumbent actors. While neither case casts a clear vote for either strategy, both show the virtues of a multifaceted approach to studying energy transition, which places climate motivations and political economies in the context of a wider set of interests and coalition building around renewable energy.

9.1 Overview of the Case Studies

In the first decade of the 2000s, both Brazil and South Africa began to contemplate the task of transforming their electricity sectors. For both, supply crises (in 2001

I would like to thank Virginia Haufler, the editors, and other chapter writers for helpful suggestions for revising the original drafts of this chapter. The chapter draws on my recent book (Hochstetler 2021) as well as other materials.

for Brazil, in 2008 for South Africa) were the proximate causes behind the shift, with new renewables like wind and solar power appearing as possible routes to diversifying the energy system for energy security despite their higher costs. Over the same decade, both also began to engage more extensively with international climate negotiations that increasingly called for comprehensive energy transitions and more commitments from emerging powers (Hochstetler 2012). By the end of the next decade, Brazil had seen an impressive rise in its installed capacity in wind power, although broader energy transformation there was patchy, as described. South Africa had made a promising start at decarbonizing its energy system in mid-decade, only to move into stalemate and even retreat.

These parallel experiences offer an excellent opportunity to look at the puzzle presented by this book. The opportunity is especially good since my recent study of the two concluded that the Brazilian case exemplified a more technical, self-reinforcing status quo process grounded in existing institutions and processes while the South African transition process faced continuous politicized scrutiny and opposition (Hochstetler 2021: 222). While the basic outcomes might seem to easily support those arguing in favor of aiming for a status quo of policy stability, this chapter argues for a more nuanced conclusion.

In particular, while the depoliticized Brazilian approach enabled specific achievements like the rapid growth of wind power, it proved unable to stimulate a wider process of energy transition, leaving Brazil's petroleum ambitions untouched (Viola and Franchini 2018: 142–146). Extending to fuller energy transition would evidently require the more overtly contentious battle with powerful actors that South Africa has not been able to avoid, given that nearly all of the latter's greenhouse gas (GHG) emissions are in the energy sector and nearly half of those in electricity. Thus, the final overarching argument is that full decarbonization faces an apparently inevitable confrontation over the distributive politics of shutting down existing and potential fossil fuel production and use (see VanDeveer, Chapter 5, this volume).

The need for confrontation does not mean the battle is inevitably lost. I argue that while decarbonization processes activate a powerful set of probable fossil fuel losers that will mobilize to resist change (see also Moe 2015), other political economies of energy transition – industrial policy, consumption issues, and siting – may create either allies or opponents for them. That balance of incentives and disincentives for transition in the multiple political economies of energy transition sets the overall national transition outcome and allows important advances in decarbonization, if not a full achievement of it.

Brazil and South Africa are important case studies for this research puzzle, as much research on climate politics has focused on the advanced industrial democracies. As emerging powers, these countries widen the evidence base for drawing conclusions about broader theoretical questions. In addition, the countries are regional powers, which often are models for their regions. Finally, much of the world's future GHG emissions is expected to come from emerging and other developing countries, so understanding climate politics there – and any differences with the Northern countries more commonly studied – is particularly important for understanding future GHG emissions trajectories. The conclusions are based on multiple periods of field research and about ninety original interviews with actors in the climate and energy sectors in the two countries. These are more fully referenced in my recent book on energy transition in Brazil and South Africa (Hochstetler 2021).

9.2 Energy Transition, Policy Stability, and Politicization: Four Political Economies

Energy systems and energy system change have wide and diverse effects on national and international political economies. That is, they reflect a series of different basic structures of interests that activate different coalitions of state and societal actors. Some of the most important are those related to climate change and decarbonization, where actors can be divided by whether they hold "climate-forcing" or "climate-vulnerable" assets (Colgan et al. 2021). But energy systems also have other important political economies. There is a productive sector that produces the energy used, with its own firms, careers, technologies, innovations, and growth trajectories. Beyond that, energy is consumed in some form by both individual households and economic actors. Finally, energy infrastructure is placed in particular communities, which may or may not welcome it. All of these political economies of energy systems are potentially subject to state regulation and control. The energy sector is one that has historically seen significant state ownership and control, which has continued in most countries even after a period of partial liberalization (Victor and Heller 2007; Victor et al. 2011). This makes energy transition a particularly relevant location to view the effects of state policy and whether policy stability or a more flexible and contention-savvy policy strategy will be more likely to build successful outcomes of energy transition.

The climate political economy of energy transition is where this theoretical debate has been put forward most explicitly (Paterson et al. 2022). On the one hand, many climate scholars argue that a status quo of policy stability is necessary because of the long time horizon of climate change and the need to provide credible signals of future policy to motivate the full range of public and private action needed. The exact logics and mechanisms behind this argument vary. From an economics or business perspective, the motivation is the need to reduce regulatory risk and to send clear and consistent signals to private economic actors (Blyth et al. 2007); such perspectives often prefer a carbon tax as mechanism. Others are more oriented to institutional or policy arrangements that are designed to deliberately lock in the desired pathway, including by creating positive feedback mechanisms (Levin et al. 2012; Lockwood et al. 2016). As a group, these approaches are looking to isolate climate policy from many current political debates and contestation.

Conversely, this same policy arena has generated some of the strongest counterarguments that decarbonization is an inherently conflictual and political process, which will belie any easy locked-in solutions on a scale that is fully adequate for the transformations necessary (Aklin and Mildenberger 2021; Paterson et al. 2022). In fact, scholars argue, efforts to produce such long-term solutions that ignore politics are likely to lock in the preferences of powerful fossil fuel actors (e.g. Brulle 2018). The fact that decarbonization ultimately requires shutting down and leaving unexploited sources of fossil fuels contributes to this conflict. Yet the strength of many current political and economic elites derives either directly or indirectly from fossil fuels. That makes climate policy and true energy transition an existential threat to very powerful actors and requires substantial state capacity to restrain and manage their power, in this view (Meckling and Nahm 2018; Moe 2015).

The dynamics of stability/politicization in the production political economy are quite different, however. Here, energy transition is often driven through the dynamics of green industrial policy, or specific state initiatives to promote renewable energy and other related sectors, such as electric vehicles (e.g. Meckling and Nahm 2019). In green industrial policy literatures, some element of policy stability is seen as necessary to motivate public and private actors to respond to the incentives given. However, the industrial policy literature also places a great deal of emphasis on the need to plan withdrawal of the incentives in a timely fashion in order to prevent excessive rents and ensure efficient and competitive development of the new industries (Pegels 2014), so even the pro-stability argument is qualified in this area. Otherwise, the production dynamic is generally one of providing positive incentives for decarbonizing action, so it can generate allies for supporting stronger climate policy and consequently has less of the inherent conflict associated with climate politics.

In the consumption political economy of energy transition, scholars ask questions about energy access, the quality of energy supply, and energy pricing levels and structures, that is, whose energy use is subsidized or cross-subsidized (Hochstetler 2021: 132–135). There is little direct discussion about policy stability versus contention, but the political economy itself is one with implications for both. Policy stability can be enforced by the physical infrastructure of distribution, with options locked in and out. For example, if energy access is provided through centralized grids (electricity) and networks (e.g. vehicle fuel distribution), then certain kinds of policy choices become much more difficult. The large scale and long lifespans of existing infrastructure make change more difficult – although these considerations also work to reinforce transitions once new infrastructure, for example electrical vehicle charging networks, is built. The physical lock-in is commonly reinforced by political lobbying by actors who benefit from the system. For example, the utilities that manage central electricity grids have often been hostile to the development of small-scale solar power and warn of utility "death spirals" if self-provision is encouraged (Stokes 2020: 103). Quality and pricing issues are rarely themselves locked in directly, but decisions on these topics are very important for creating allies or opponents for energy transition. Consumer subsidies for fossil fuels, for example, are often defended with mobilized protests (e.g. Akanle et al. 2014) that provide a powerful force against leaving fossil fuels behind.

Finally, the political economy of energy transition siting – where fuels and industrial inputs are sourced, where energy is refined and produced, where the built infrastructures of electricity and pipelines are placed – share some of the features of the consumption political economy. The physical facilities themselves lock in particular energy choices and particular locations of impact, while the prospect of new ones, necessary for any transition to take place, often face significant resistance from prospective host communities (McAdam and Boudet 2012). Whether hosting an energy-related facility brings benefits like jobs and infrastructure to host communities (e.g. Xia and Song 2017) or imposes significant costs on often vulnerable hosts (e.g. Avila 2018) is a matter of considerable theoretical debate. This is in part because the costs and benefits genuinely vary. The specific balance of costs and benefits is, again, likely to create opponents and proponents of energy transition. The solutions to those conflicts, to build or not to build, will then again physically reinforce any energy choice.

This discussion has remained quite abstract, while pointing out at numerous points that there is likely to be a fair amount of variation and specificity in national energy transition efforts. In addition, the final mix of the political economies will also matter. Thus, even policy and institutional consistency on climate change itself, narrowly defined, may be undermined – or reinforced – by decisions taken by other actors on, say, who has access to subsidized finance for building infrastructure (a production question). As a result, it is helpful to examine two vignettes of how energy transition debates unfolded in Brazil on the new alternatives of wind and solar power, and the simultaneous development of fossil fuel production in the oil and gas sector. This will be followed with a brief look at South Africa. Collectively, they will allow me to sketch how the political economies of energy transition can inform the broader question of this book.

9.3 Energy Transition in Brazil: Advancing the New Renewables

Energy transition in Brazil looks quite advanced, especially if one focuses on the growth of alternative renewables like wind power and (more recently) solar

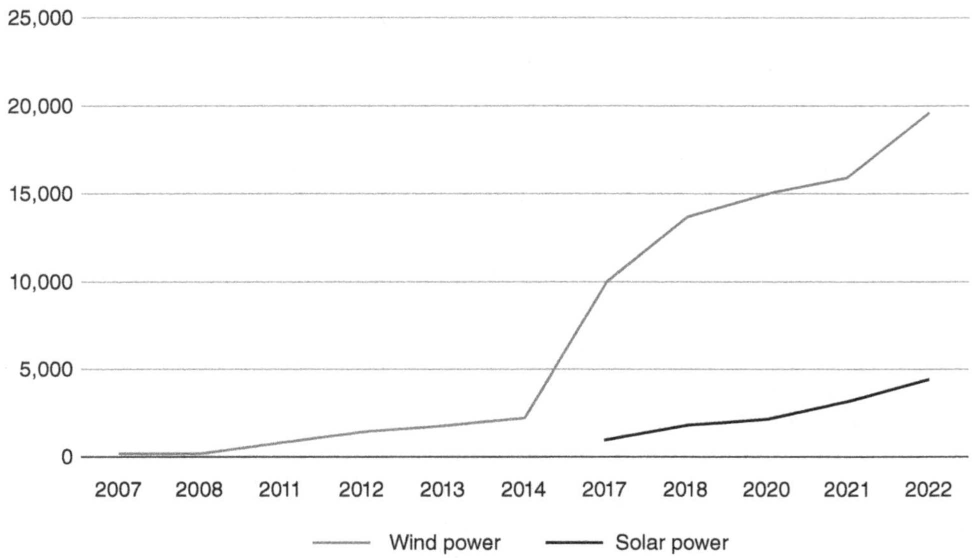

Figure 9.1 Installed MW of wind and solar power in Brazil, 2007–2022. Compiled by author from data from the Empresa de Pesquisa Energêtica at https://dashboard.epe.gov.br/apps/anuario-livro/#22_Capacidade_instalada_por_fonte_(MW) [dashboard.epe.gov.br]

power (see Figure 9.1). Brazil has traditionally used hydroelectric power, which has fewer GHG emissions but more socio-environmental conflict than fossil fuel sources of electricity. Instead, almost all of its emissions have historically come from deforestation, agriculture, and other land-use sources of greenhouse gases (Basso 2019: 5). Less than 3 percent of its GHG emissions come from electricity itself (Ministério de Minas e Energia 2018: 245). This means that the contentious debates there around climate change have rarely focused on energy or electricity, making it a country almost uniquely able to address climate change without engaging in a politicized debate around energy transition. In fact, most of Brazilian decision-making around renewable energy has in fact come in depoliticized, steady, standard operating-procedure bureaucratic decisions (Hochstetler 2021: 229–230).

Figure 9.2 summarizes ten-year electricity planning in Brazil, showing the smooth and consistent expansion of wind and (eventually) solar power that resulted. Figure 9.2 shows why the idea of depoliticized, stable decision-making on climate is so appealing. The unfolding of wind power in Brazil exemplifies what climate activists hope for. Yet, while wind power expanded steadily after 2009 in Brazil, solar power lagged; the first auctions for the right to build solar power were held only in 2014 and it is only beginning to really take off a decade later. This difference cannot be explained by a narrow focus on climate change, since there are no

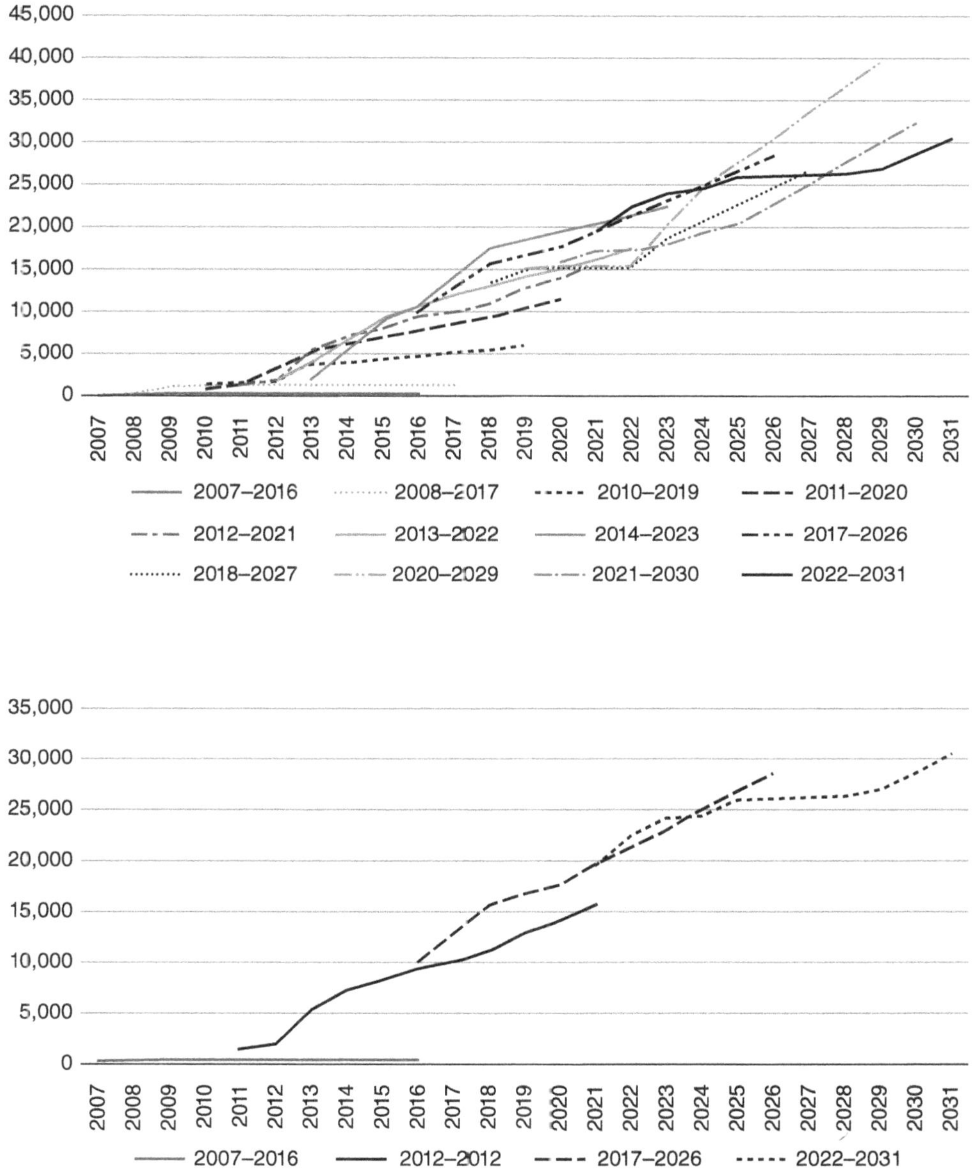

Figure 9.2 Ten-year electricity planning in Brazil: (a) ten-year wind power projections in Brazil, in installed MW and (b) ten-year solar power projections in Brazil, in installed MW.
Source: Compiled by author from data from the *Empresa de Pesquisa Energética* (2024)
Note: Starting points for each series are actual contracted/built quantities for that year.

climate-based reasons to avoid solar power in a sunny, tropical country. Instead, the different fates of wind and solar power in Brazil came down to the cost/consumption and industrial policy political economies of energy transition.

Cost considerations – important for consumption decisions – are a significant part of the story. As Brazil first began to consider building the new renewables in 2001, solar power was much more expensive than wind power and other alternatives. While wind power was also more expensive than non-solar alternatives, Brazilian energy modelers believed the higher costs could be compensated with the greater potential to build an industrial sector that would supply components for wind power (Hochstetler 2021: 93).

Existing industrial capacities were further leveraged with routine policies in the Brazilian National Development Bank (BNDES) that provided subsidized funding for projects that included locally produced inputs, allowing them to win Brazil's cost-based auctions to supply electricity to the grid. The result for wind power was the classic green spiral where industry and government worked together in a self-reinforcing way to steadily expand the sector (Hochstetler 2021: 103–104). Conversely, industrial development was much more difficult for solar power and BNDES finance was less decisive, so it lacked this self-reinforcing mechanism. Solar power did not expand until the sharp fall in the global cost of solar components took place after 2011. This contrast between wind and solar power therefore supports the versions of the policy stability model that emphasize path dependence, especially when reinforced through the building of new actors and coalitions in favor of the direction of transition.

As a final part of the story, the siting dimension of energy transition in Brazil worked against the dynamic established in the other political economies, although it was not decisive. About one-quarter of local host communities (nineteen of seventy-seven) strongly objected to and resisted the building of wind power farms in their communities. This resistance was most likely in communities where land tenure was uncertain, as in historic former slave communities where residents had had informal access to beaches and other locations without formal ownership of them (Hochstetler 2021: 202–205). The resistance was strong enough to generate a policy change away from a permissive environmental licensing process that had assumed there would be no negative local impacts (Hochstetler 2021: 191). Ultimately, the resistance was unable to block wind power and was more influential over time in determining exactly where wind power would be built in Brazil's many potential sites. In contrast, there were no community mobilizations against hosting solar power, but that did not prove to be a decisive advantage for solar power given the other political economies of energy transition. It is worth noting that energy siting in Brazil is generally a highly contentious process, with communities and activists poised to quickly mobilize against new large hydropower sites – although fossil fuel-based electricity plants are the least likely to generate host community opposition (Hochstetler and Tranjan 2016).

9.4 Energy Non-transition in Brazil: Advancing the Oil and Gas Sector Too

All of Brazil's gains in renewable energy production have not limited the continuing expansion of fossil fuels. Instead, the renewables and fossil fuels expanded together, following the discovery of large deposits of oil and gas deep offshore in 2007. Other scholars have noted this counterintuitive result (e.g. Aamodt 2018; Basso 2019; Viola and Franchesi 2018). In this chapter, I show that both outcomes result from very similar policy processes, with the exception of the climate political economy. With existing institutions repurposed to add more new renewables to the electricity sector without taking on a transformative climate ambition in the energy sector, Brazil is actually moving away from true climate-friendly energy transition even as its wind and solar power capacity continues to grow.

As noted, the climate political economy is the one that makes most explicit the demand that existing fossil fuel production must be truncated and even shut down to meet its end goals (Hochstetler 2021: 30). Unlike many countries with oil and gas production, Brazil does not have a long-standing dependency on the sector for state revenues and export earnings. Its historic reserves were small, and the available rents followed suit. Neither its fossil fuel subsidies nor the share of petroleum rents have surpassed 3 percent of gross domestic product (GDP).[1] Yet the state-controlled oil company, Petrobras, has long been "the most emblematic state-owned enterprise" (Mortari et al. 2021: 147), a "jewel in the crown" of both Brazil's domestic and international politics (Ceppi and Doval 2020: 102), with all of the cultural and economic power of other statist fossil fuel incumbents. Petrobras's many deep ties to political and economic elites became evident when it became one of the primary targets of a large corruption scheme in the construction sector, known as *Lava Jato* or Carwash, eventually paying $853.2 million in fines to Brazil and the United States (Viswanatha and Lewis 2018).

Neither Brazil's 2016 nationally determined contribution (see Allan, Chapter 14, this volume) nor subsequent revisions even mention the fact that Brazil is an increasingly important oil and gas producer. They note instead that the renewables share in Brazil's energy matrix was already three times the world average (Brazil Second Revised NDC 2022) and that they expect the share to rise by expanding non-hydro renewables. Governmental narratives on oil production and climate change either stress the way that petroleum expansion can pave the way for energy transition or assert that Brazil has a right to develop its resources in a world where most GHG emissions are associated with the Global North (Víglio et al. 2019: 146, 148). The fact that deforestation and land-use change dominate Brazilian GHG

[1] See, respectively, the IMF Climate Change Indicators Dashboard, "Fossil Fuel Subsidies" dataset: https://climatedata.imf.org/datasets/d48cfd2124954fb0900cef95f2db2724/explore; and World Bank Group, "Oil Rents (% of GDP)" dataset: https://data.worldbank.org/indicator/NY.GDP.PETR.RT.ZS?name_desc=false.

emissions allows even actors serious about climate action to elide Brazil's fossil fuel industry. Yet political conflict also intervened in this result. The Climate Law passed by the National Congress in 2009 included clauses that called for phasing out fossil fuels and replacing them with renewables but then (and current) President Lula da Silva used his line-item veto to remove them. The veto followed the wishes of the Ministry of Mines and Energy as well as Lula's presidential successors (Aamodt 2018: 386).

The regular ten-year energy planning documents that lay out steady expansion for wind and now solar power also contain chapters on oil and gas production. But while the former discussions frequently mention Brazil's climate commitments, the latter do not mention climate change at all. They simply register amounts of reserves, rates of production and reserve depletion, and clearly are oriented toward maximizing possible production levels in the sector (e.g. Ministério de Minas e Energia 2023). In this chapter, but not in the rest of the document, sustainability means sustained production. The same sophisticated energy planning agency using its same standard optimization models bolstered the expansion of wind and solar power and also planned to boost oil and gas production. The planning logics are consistent on most metrics, favoring factors like low prices and energy security, but have opposite implications for decarbonization.

Similarly, BNDES used very similar decision-making metrics to support both wind power and oil and gas expansion. By statute, BNDES values Brazilian long-term economic development and both short- and long-term job creation, with a frequent focus on national industrial development. As seen in the last section, the local content conditions for its loans helped to promote the development of wind, but not solar, power, by providing an additional incentive and political support for wind power development through incentives for supply chain development. For oil and gas, BNDES's loans were even more decisive for Petrobras's investment needs after the pre-salt discovery. These coincided in time with the global financial crisis and a temporary disappearance of private investment funds. BNDES's loans to Petrobras rose by more than five times (R$ 8.8 billion to R$ 47.87 billion) between 2007 and 2012. In that final year, loans to Petrobras were 30.69 percent of BNDES's loan portfolio (Barbosa 2013: 31), showing the importance of the petroleum parastatal to this public bank. The bank's local content requirements were buttressed by an explicit governmental policy for oil and gas inputs, again providing increased political and economic incentives for the sector's expansion (Mortari et al. 2021). BNDES's public database of all its lending shows no funds for extraction after 2013 and only loans for tankers and ports after 2014.[2] New public funds for Petrobras disappeared with austerity, the *Lava Jato* corruption scan-

[2] See BNDES database, www.bndes.gov.br/wps/portal/site/home/transparencia/centraldedownloads#.

dal, and the Workers' Party's exit from the presidency, but BNDES still provided irreplaceable support for Petrobras's expansion into deep sea production at a key moment that might otherwise have seen the initiative fail or develop more slowly.

The consumption political economy proved to be a double-edged sword for the oil and gas sector. The national government continues to control a majority of Petrobras's shares and has frequently intervened in the company to pursue other national goals. Some of the most notable policies came in Dilma Rousseff's first government (2011–2014) when Petrobras was forced to sell its products domestically at prices below what it paid for them internationally. Taxes were also lifted for automobiles as well as for oil and gas (Basso 2019: 10). The policy was damaging to Petrobras as a company, but the subsidized prices also provided a climate-hostile subsidy for fossil fuel consumption. Gasoline was used at the expense of biofuels in Brazil's common flex-fuel cars, for example (Basso 2019: 10).

For local communities that host oil and gas installations, debates about local impacts have largely centered on the royalties long distributed to producing municipalities. These provided at least 5 percent of municipal budgets for 122 communities over a five-year period (Lima-de-Oliveira and Alonso 2017: 581). While the development outcomes for these municipalities did not improve (Lima-de-Oliveira and Alonso 2017: 584), they fought hard to retain the royalties through several contentious congressional debates (Trojbicz 2019). Alongside the political debates among elected elites, large crowds marched in Rio de Janeiro and other cities to demand continued royalties, even as there was a striking absence of contestation around the oil spills and other environmental hazards around offshore oil production (Hochstetler 2011). In other words, the question of local impacts spilled outside of standard bureaucratic processes in several ways, with both congressional and street mobilization dominated by those seeking continued oil and gas production.[3]

In general, the political economies of the oil and gas sector (and the government policies that shaped them) thus worked to reinforce that sector even as wind and solar power expanded. Many of the same institutions – BNDES, the Energy Planning Agency, the climate institutions – promoted both, either intentionally or by omission, in the case of the climate institutions. Thus, the final summary of the Brazilian case shows that a status quo of bureaucratic politics *can* promote important changes in the energy sector but did not transform the sector as a whole. Contentiousness was generally low across the board but emerged in 2010 over the issue of adding a phaseout of oil and gas to the National Climate Law. Very new dynamics within Lula's third presidency (2023–2027), beyond the scope of this

[3] NGOs like those in the Climate Observatory (*Observatorio do Clima*: www.oc.eco.br) have raised the topic of the local and global environmental costs of oil and gas production in Brazil, but local communities have not taken them up.

chapter, also show that efforts to actually block the expansion of oil and gas sectors are introducing the politicization and conflict that had been avoided.

9.5 Energy Non-transition in South Africa

If South Africa is to reduce its GHG emissions, it can only do so by decarbonizing its energy sector, responsible for about 80 percent of its emissions (Hochstetler 2021: 36). However, it has the classic scenario expected by the authors who say transitions can only be politicized: The whole South African economy has been organized around a "minerals-energy complex" of political and economic elites who have relied on historically cheap coal-based electricity (Fine and Rustomjee 1996). For the parastatal Eskom and heavy electricity consumers in the Energy Intensive Users Group (EIUG), whose twenty-eight firms use 40 percent of South Africa's electricity supply, the prospect of energy transition appeared as an existential threat from when it was first proposed in 1998. Early on, Eskom was able to simply block requests to diversify from coal, but after the crippling supply shortages after 2008, the sector quickly became the site of significant contestation, whether at consultative meetings on climate in 2009 or in debates around planning South Africa's long-term electricity procurement plans (Hochstetler 2021: 39–57). Figure 9.3 illustrates the country's inability to ever settle into a clear vision of

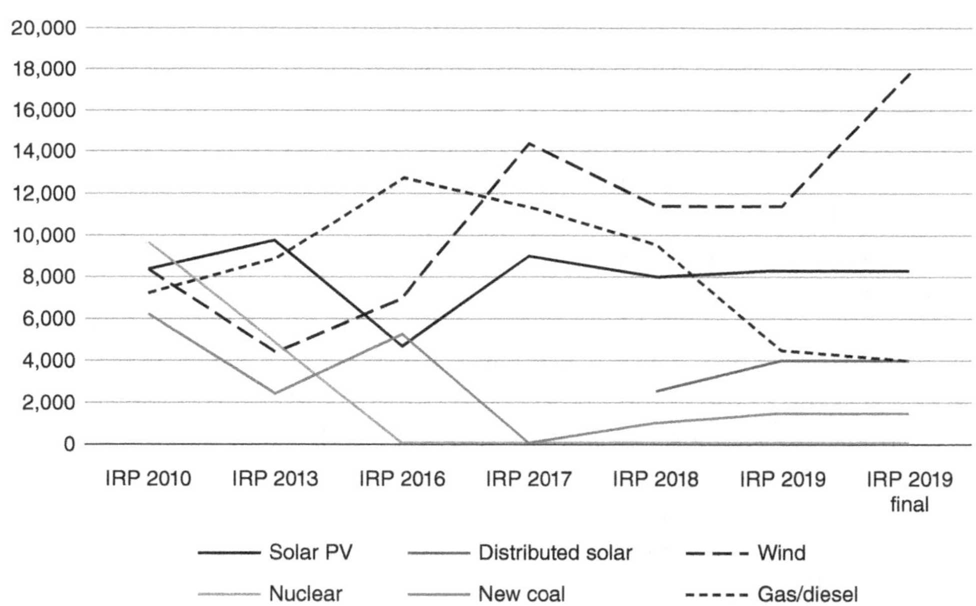

Figure 9.3 Projected total electricity procurement in installed MW to 2030 by fuel type in drafts of the South African Integrated Resource Plan for Electricity. Source: Hochstetler (2021): 47.

its electricity sector, with successive plans for electricity procurement fluctuating wildly as different actors gained short-term influence in the planning process.

Institutions followed suit, with the 2009 Inter-Ministerial Committee on Climate Change omitting the most powerful relevant state actors, the Ministry of Energy, the Treasury, and Eskom, leaving a weak Ministry of Environmental Affairs unable to lead decisive climate action among its partners (Tyler and Hochstetler 2021: S191). In short, while effective climate policy might require policy stability and a steady hand, the South African experience shows that putting such policies and institutions in place is a daunting requirement in a context where there are powerful interests in blocking transition. There has been very little evidence that the South African state has the capacity to do so in the face of "the presence of a strong, well-resourced actor like Eskom fighting an existential threat [that] clearly drove the emergence of these counter-coalitions and their battles" (Hochstetler 2021: 224). Yet, while the basic South African climate story looks like the inevitable politicization following an activated fossil fuel power, the other dimensions of energy transition there show places where more harmonious and steady pro-transition coalitions and policies might have been built.

As already noted, the industrial policy and sectoral growth dimensions of energy transition have often provided a positive motivation for change, even resulting in a self-reinforcing green spiral of industry growth and governmental reinforcement for many early adopters (Zysman and Huberty 2014). South African policymakers were very interested in this possibility, with numerous government departments putting forward plans (Hochstetler 2021: 112–115). This opportunity was interrupted in South Africa because reformers made the renewable energy program the entry point for private generation, which introduced the contentious and further politicized dynamics of sectoral privatization to decarbonization. This was in part because Eskom had earlier refused to decarbonize itself, but it eventually led to a poisonous association of public generation with coal and nuclear power while nearly all wind and solar power generation was private. This magnified Eskom's opposition to shifting fuels and helped to shift the labor movement – supportive of climate action but ideologically opposed to private generation – into the opposition to developing wind and solar power (Hochstetler 2021: 115–117). The fits and starts visible in Figure 9.3 further drove away much private interest in localizing input manufacturing so that most jobs associated with renewable energy held little interest for labor, completing a negative rather than positive spiral.

Similarly, South African electricity planners had hoped that solar power could finally bring electrical service to all South Africans, especially the Black South Africans unserved by apartheid and still disproportionately unlinked to the national grid. Many of those disconnected rejected the minimal household service offered through solar power, however, with some researchers characterizing the offer as environmental injustice (Monyei et al. 2018). Instead, wind and solar power were most embraced by wealthier consumers opting for self-provision off the grid. The EIUG also appreciated the steadily dropping prices of renewables across the four auctions as coal prices simultaneously rose (Hochstetler 2021: 27). Thus, while this political economy can ease politicization and conflict, it did not in South Africa, with its racialized history of inequitable electricity service provision.

The South African government designed its renewable energy procurement program in ways that might have built opposition or support for broader energy transition in the local communities that host electricity infrastructure. On the one hand, it used the auction procurement mechanism for electricity, which tends to provide incentives for cheaper electricity – itself a source of support for transition – through larger and more intrusive installations that are more likely to encounter opposition from host sites (Grashof 2019). On the other hand, bidders won the auctions to build wind and solar farms based on bids that considered not only the price offered but also the benefits they promised to communities within 50 km of an installation. These included mandatory community ownership up to 5 percent, as well as promises for jobs for locals and for Black South Africans in particular (Tait et al. 2013). In the end, only one community in South Africa had substantial mobilization against wind power (Hochstetler 2021: 210), but the small number of projects procured and built largely neutralized both the potential positive and negative benefits of renewable energy for local committees.

In summary, the South African experience demonstrates the basic dynamic of the politicized scenario. In a country with powerful actors committed to fossil fuels, politicization of energy transition was unavoidable. Those powerful actors, led by the parastatal Eskom, successfully outright blocked and maneuvered behind the scenes against energy transition for more than two decades. At the same time, the South African experience includes several places where there were points of intervention – actual and potential – that could have led to other outcomes, either less or more of an energy transition than actually happened. Had South Africans found a different way around the equally thorny politicization there around privatization, for example, transition might have been smoothed. Thus, its lesson is that politicization may be unavoidable in some circumstances but still leaves multiple potential leverage points that may affect outcomes.

9.6 Conclusions

Does meaningful decarbonization depend on policy stability that makes climate policies and institutional development irreversible, or does it depend on mastering a messy political conflict with steps forward and backward that might be inherent in large-scale political economy transitions? The case studies of energy transition in Brazil and South Africa that are considered here ultimately conclude that messy political conflict is probably an inevitable step in a full-scale energy transition in countries that produce fossil fuels. Fossil fuel producers and those allied with them politically and economically can see the existential threat that decarboniza-tion holds for them and have the resources to fight that end. While this conclusion cannot be avoided, I would argue that the case studies here provide considerable nuance around it and even show important contributions that policy stability can make to the final outcome. Many of these nuances are best seen in considering other non-climate political economies around energy transition, which can skirt the direct threat that climate action poses to the fossil fuel coalitions.

The experiences of the two countries show that the potential for green industry developments can provide powerful short- and long-term counterweights to the fossil fuel coalition. Brazil shows this positively, as both public and private actors worked together in a mutually reinforcing green spiral to support ever-larger quantities of wind power production there; the absence of such a coalition slowed the introduction of solar power. South African shows it negatively, where policy instability in the form of electricity planning sent discouraging signals to private investors and installation firms, turning them away from the South African economy. Even though the expansion of Brazilian wind power was not enough to transform the whole energy sector, decarbonization requires having an alternative ready to replace the fossil fuels and Brazil shows how a state–society coalition can get that done. The consumption and siting political economies of energy transition are other dimensions where other allies can be developed as counterweights to the fossil fuel coalitions. It is noteworthy that this is not a process heavily shaped by NGOs in either country, although they have strong NGO sectors.

The discussion here sketches developments over two decades that also show that policy stability and politicization can both dominate at different points in time in the story. Brazil manages quite a bit of policy consistency in most political economies for most of those decades, but politicization emerges at key moments when fossil fuel producers are explicitly threatened, as with the 2009 climate legislation. For South Africa, the threats and politicization are more present most of the time, but there are also clear points where predictable policy interventions could have diffused some of the conflict, notably around privatization. Thus, the two approaches may necessarily appear together rather than being as clearly opposed as the literature can suggest.

References

Aamodt, S. (2018). Environmental ministries as climate policy drivers: Comparing Brazil and India. *Journal of Environment and Development* 27(4), 355–381. https://doi.org/10.1177/1070496518791221

Akanle, O., Adebayo, K., and Adetayo, O. (2014). Fuel subsidy in Nigeria: Contexts of governance and social protests. *International Journal of Sociology and Public Policy* 34(1/2), 88–106.

Aklin, M., and Mildenberger, M. (2020). Prisoners of the wrong dilemma: Why distributive conflict, not collective action, characterizes the politics of climate change. *Global Environmental Politics* 20(4), 4–27.

Avila, S. (2018). Environmental justice and the expanding geography of wind power conflicts. *Sustainability Science* 13, 599–616.

Barbosa, P. (2013). *O Endividamento da Petrobras com o BNDES no Periódo Pós-2008 e Impactos Contábeis e Econômico-Financeiros. FGV-IBRE, Texto de Discussão No. 36.* Rio de Janeiro: FGV-IBRE.

Basso, L. (2019). Brazilian energy-related climate (in)action and the challenge of deep decarbonization. *Revista Brasileira de Política Internacional* 62(2), e2.

Blyth, W., Bradley, R., Bunn, D. et al. (2007). Investment risks under uncertain climate change policy. *Energy Policy* 35(11), 5766–5773.

Brazil. (2022). Federative Republic of Brazil: Paris Agreement Nationally Determined Contribution (Second Revised), March 21. https://unfccc.int/sites/default/files/NDC/2022-06/Updated%20-%20First%20NDC%20-%20%20FINAL%20-%20PDF.pdf

Brulle, R. J. (2018). The climate lobby: A sectoral analysis of lobbying spending on climate change in the USA, 2000 to 2016. *Climatic Change* 149(3), 289–303.

Ceppi, N., and Pereyra Doval, G. (2020). Petrobras entre Nacionalistas e Liberales: del PT a Bolsonaro. *Revista de Relaciones Internacionales, Estrategia e Seguridad* 15(1), 99–115.

Colgan, J. D., Green, J. F., and Hale, T. N. (2021). Asset revaluation and the existential politics of climate change. *International Organization* 75(2), 586–610.

Empresa de Pesquisa Energêtica (2024). Planos Decineis de Expansão de Energia. www.epe.gov.br/pt/publicacoes-dados-abertos/publicacoes/plano-decenal-de-expansao-de-energia-pde

Fine, B., and Rustomjee, Z. (1996). *The Political Economy of South Africa: From Minerals Complex to Industrialization.* London: C. Hurst & Co.

Grashof, K. (2019). Are auctions likely to deter community wind projects? And would this be problematic? *Energy Policy* 125, 20–32.

Hochstetler, K. (2011). The politics of environmental licensing: Energy projects of the past and future in Brazil. *Studies in Comparative International Development* 46(4), 349–371.

Hochstetler, K. (2012). Climate rights and obligations for emerging states: The cases of Brazil and South Africa. *Social Research* 79(4), 957–982.

Hochstetler, K. (2021). *Political Economies of Energy Transition: Wind and Solar Power in Brazil and South Africa.* Cambridge: Cambridge University Press.

Hochstetler, K., and Tranjan, J. R. (2016). Environment and consultation in the Brazilian democratic developmental state. *Comparative Politics* 15(3), 74–94.

Levin, K., Cashore, B., Bernstein, S., and Auld, G. (2012). Overcoming the tragedy of super wicked problems: Constraining our future selves to ameliorate global climate change. *Policy Sciences* 45(2), 123–152. https://doi.org/10.1007/s11077-012-9151-0

Lima-de-Oliveira, R., and Alonso, M. L. (2017). Fueling development? Assessing the impact of oil and soybean wealth on municipalities in Brazil. *The Extractive Industries and Society* 4, 576–585.

Lockwood, M., Kuzemko, C., Michell, C., and Hoggett, R. (2016). Historical institutionalism and the politics of sustainable energy transitions: A research agenda. *Environment and Planning C: Politics and Space* 35(2), 312–333.

McAdam, D., and Shaffer Boudet, H. (2012). *Putting Social Movements in Their Place: Explaining Opposition to Energy Projects in the United States, 2000–2005.* Cambridge: Cambridge University Press.

Meckling, J., and Nahm. J. (2018). The power of process: State capacity and climate policy. *Governance* 31, 741–757.

Meckling, J., and Nahm. J. (2019). The politics of technology bans: Industrial policy competition and green goals for the auto industry. *Energy Policy* 126, 470–479.

Ministério de Minas e Energia. (2018). Plano Decinal de Expansão de Energia 2027 [Draft for Comment]. Brasília: Ministério de Minas e Energia and Empresa de Pesquisa Energêtica.

Ministério de Minas e Energia. (2023). Plano Decinal de Expansão de Energia 2032. Brasília: Ministério de Minas e Energia and Empresa de Pesquisa Energêtica.

Moe, E. (2015). *Renewable Energy Transformation or Fossil Fuels Backlash: Vested Interests in the Political Economy*. London: Palgrave.

Monyei, C. G., Adewumi, A. O., and Jenkins, K. E. H. (2018). Energy (in)justice in off-grid rural electrification policy: South Africa in focus. *Energy Research & Social Science* 44, 152–171.

Mortari, V. S., Garcia Ribeiro, C., Loural, M., and Oliveira, A. (2021). A Petrobras como Agente do Desenvolvimento Produtiva no Brasil: Importância e Esvazamento. Revista da Sociedade Brasileira de Economia Política 61, 143–176.

Paterson, M., Tobin, P., and VanDeveer, S. D., (2002). Forum: Climate governance antagonisms: Policy stability and repoliticization. *Global Environmental Politics* 22(2), 1–11.

Pegels, A. ed. (2014). *Green Industrial Policy in Emerging Countries*. London: Routledge.

Stokes, L. C. (2020). *Short-Circuiting Policy: Interest Groups and the Battle over Clean Energy in the States*. Oxford: Oxford University Press.

Tait, L., Wlokas, H. L., and Garside, B. (2013). *Making Communities Count: Maximizing Local Benefit Potential in South Africa's Renewable Energy Independent Power Producer Procurement Programme*. London: International Institute for Environment and Development.

Trojbicz, B. (2019). Federalism and governability in Brazil: Oil royalties in dispute. *Bulletin of Latin American Research* 38(5), 607–623.

Victor, D., and Heller, T. C. eds. (2007). *The Political Economy of Power Sector Reform: The Experiences of Five Major Developing Countries*. Cambridge: Cambridge University Press.

Victor, D. C., Hults, D. R., and Thurber, M. C. eds. (2011). *Oil and Governance: State-Owned Enterprises and the World Energy Supply*. Cambridge: Cambridge University Press.

Vígilio, J. E., Di Giulio, G. M., Barbi, F., and da Costa Ferreira, L. (2019). Narrativas Científicas Sobre Petróleo e as Mudanças do Clima e suas Reverberações na Política Climática. *Sociologias* 21(51), 124–158.

Viola, E., and Franchini, M. (2018). *Brazil and Climate Change: Beyond the Amazon*. New York: Routledge.

Viswanatha, A., and Lewis, J. T. (2018). Petrobras to pay $853.2 million to settle corruption investigations in US, Brazil. *Wall Street Journal*, September 27.

Xia, F., and Song, F. (2017). Evaluating the economic impact of wind power development on local economies in China. *Energy Policy* 110, 263–270.

Zysman, J., and Huberty, M., eds. (2014). *Can Green Sustain Growth? From the Religion to the Reality of Sustainable Prosperity*. Stanford: Stanford University Press. https://doi.org/10.2307/j.ctvqr1czp

10

Steering Political Conflicts for Climate Stability

The Case of China

YIXIAN SUN, WEI SHEN, AND JOANNA I. LEWIS

How can countries overcome challenges to achieve radical sociotechnical transformation in response to the climate crisis? Many researchers have considered possible pathways to rapid decarbonization and expressed different views: Some underscore the necessity to generate path-dependent policy processes to break carbon lock-in through entrenched and increasing support (Bernstein and Hoffmann 2019; Levin et al. 2012; Rosenbloom et al. 2019), but others indicate the need for political conflicts to challenge the power of incumbents who want to delay or prevent transition (Colgan et al. 2021; Paterson 2021a; Stokes 2020). However, these insights on climate politics and the resulting debate on policy stability versus (re)politicization have been largely drawn from cases with similar contexts, namely liberal democracies in the Global North. This bias leaves research on developing countries and emerging economies underdeveloped despite the critical role of these countries in global climate governance.

Our study recognizes the diversity of political economies across countries and the importance of contextual factors in shaping conditions for transition, and suggests that policy stability and political conflicts are not always incompatible, especially in non-Western contexts without liberal democracy (Roberts et al. 2018). We argue that policymakers can strategically make institutional arrangements to ensure durable policies that provide continuous support for low-carbon transition – what Chapter 1 refers to as "stability as policy lock-in" as well as "stability as the status quo." In other words, when windows of opportunities remain open, powerful actors can design and promote policies with strong feedback effects to weaken anti-transition interests and strengthen pro-transition ones. Such situations are likely to occur when pro-climate state actors with a strong influence on the market can shift the interest of powerful incumbent actors toward decarbonization.

We use the case of China's climate response to illustrate our argument. China is a crucial case due to its double role as the world's largest greenhouse gas emitter and largest energy consumer and producer. At the same time, the Chinese case can contribute to the development of theories about climate politics by shedding light on the possibility of managing political conflict to support policy durability in a political economy context that significantly diverges from Western-centric narratives. In this way, we demonstrate how leaders avoid politicization given the potential risk for political conflict and instead look for ways to secure stability.[1] We illustrate this through an empirical analysis that focuses on the reforms of Chinese state-owned power producers and shows how the Chinese government has strategically regrouped fossil fuel and renewable producers to manage potential political conflict, and accordingly, ensure the durability of the country's clean energy transition.

In the rest of our chapter, we present a brief overview of China's climate governance, including recent targets and policies, along with different interpretations of China's governance model. We then turn to a discussion of the role of the state in managing potential conflicts in the processes of low-carbon transition, viewed through reforms targeting China's power sector. We argue that the Chinese state has attempted to shift the interests of big state-owned power producers by regrouping them with rising renewable producers. The China case thus contrasts with South Africa (see Hochstetler, Chapter 9, this volume) where the state-owned power sector has attempted to crowd out renewable providers. In this way, our chapter is a key contribution to the debate on the role of state actors in pursuing policy stability versus politicization. We demonstrate how, in the China case, politization is in fact state-led, in contrast to what we see in many other countries. Furthermore, while most studies of climate governance in China focus on the role of the central government, China's climate response is not monolithic, and here we demonstrate the role of state-owned enterprises in the power sector. Based on this empirical case, we consider possible conditions under which the Chinese state can manage potential conflicts in transition. We conclude by discussing pathways and remaining challenges to reconciling policy stability and political conflict in the low-carbon transition.

10.1 Governing Climate Change in China: Challenges and Opportunities

As the world's most populous country and largest emerging economy, China's climate policy has large implications for the global effort to combat climate change. Since the 1990s, rapid development has not only transformed the Chinese economy and society but also made the country's energy system increasingly carbon-intensive.

[1] We are not arguing that this can only happen in China, but in China's authoritarian context, this way of managing conflicts can happen more easily.

When countries began to negotiate the United Nations Framework Convention on Climate Change (UNFCCC) in 1990, China's carbon emissions represented only 11 percent of global emissions and less than half of the emissions of the United States (Ritchie et al. 2020). Accordingly, international pressure on China for emissions reductions remained limited and China was defined as a "non-Annex I" country in the Convention (see Allan, Chapter 14, this volume). However, since then, China's emissions have grown dramatically, especially between 2000 and 2012, with an average annual increase rate of 9.5 percent (Sandalow et al. 2022). In 2006, the country surpassed the United States to become the world's largest emitter, and today, China's carbon emissions account for more than 30 percent of the global total (see Figure 10.1). Although per capita emissions in China remain much lower than those in developed countries, improvements in the standards of living in the country have led Chinese people to continuously increase their carbon footprint, which is now above the global average (see Figure 10.2).

Hence, if China cannot take strong action to quickly decarbonize its economy, the chance of reaching the 2015 Paris Agreement's goal is small. To date, China's response to climate change has received a mixed reaction as it has been seen by some as a villain responsible for worsening climate change and by others as a champion of climate action, due to its rapid development of low-carbon technology. In 2020, China announced the ambitious goal of peaking its emissions before 2030 and achieving carbon neutrality by 2060. How has China responded to climate change, and to what extent is the country on track for a rapid and deep transition toward net zero? In what follows, we briefly review China's policies and consider the key features of China's climate governance model.

10.1.1 China's Support for Low-Carbon Technologies and Dependence on Coal

As China has paid increasing attention to climate change, a major concern for its central government is to balance economic development and environmental protection (Qi and Wu 2013). In other words, climate action should not undermine the country's continuous growth. In fact, for a very long time the most important indicator to measure performance of governments at different levels in China has been targets on economic growth, and such challenges to maintaining growth may disincentivize and delay Chinese policymakers in taking strong action to combat climate change. At the same time, being considered a developing country by the UNFCCC, China has put strong emphasis on "common but differentiated responsibility" and asked for the support of developed countries for decarbonization.

Since the entry into force of the Kyoto Protocol in the early 2000s, China quickly became the largest beneficiary of the Clean Development Mechanism (CDM) by

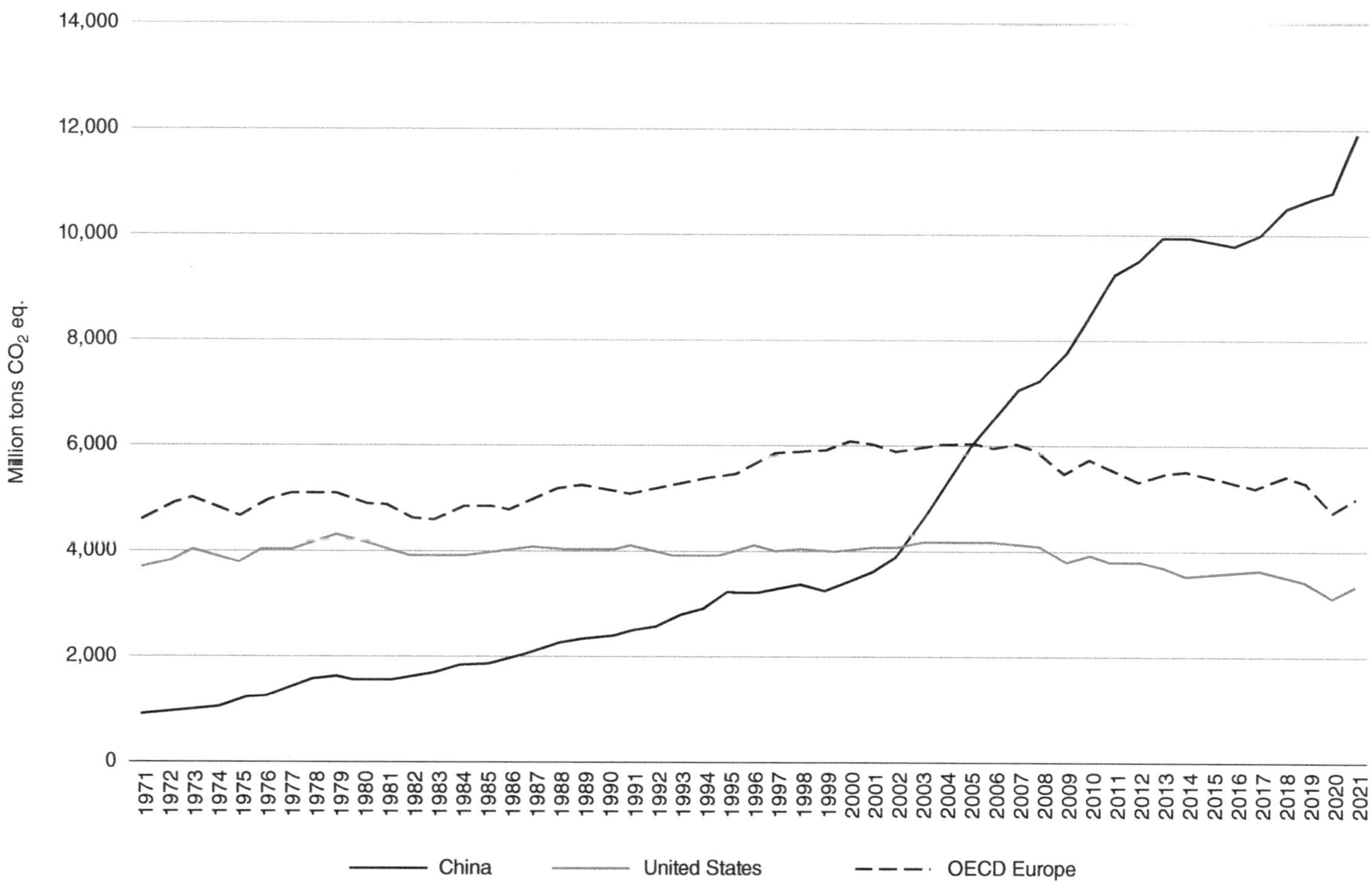

Figure 10.1 GHG emissions from energy since 1971: China, United States, and OECD Europe.
Source: GHG emissions from energy from IEA Global Energy Review.

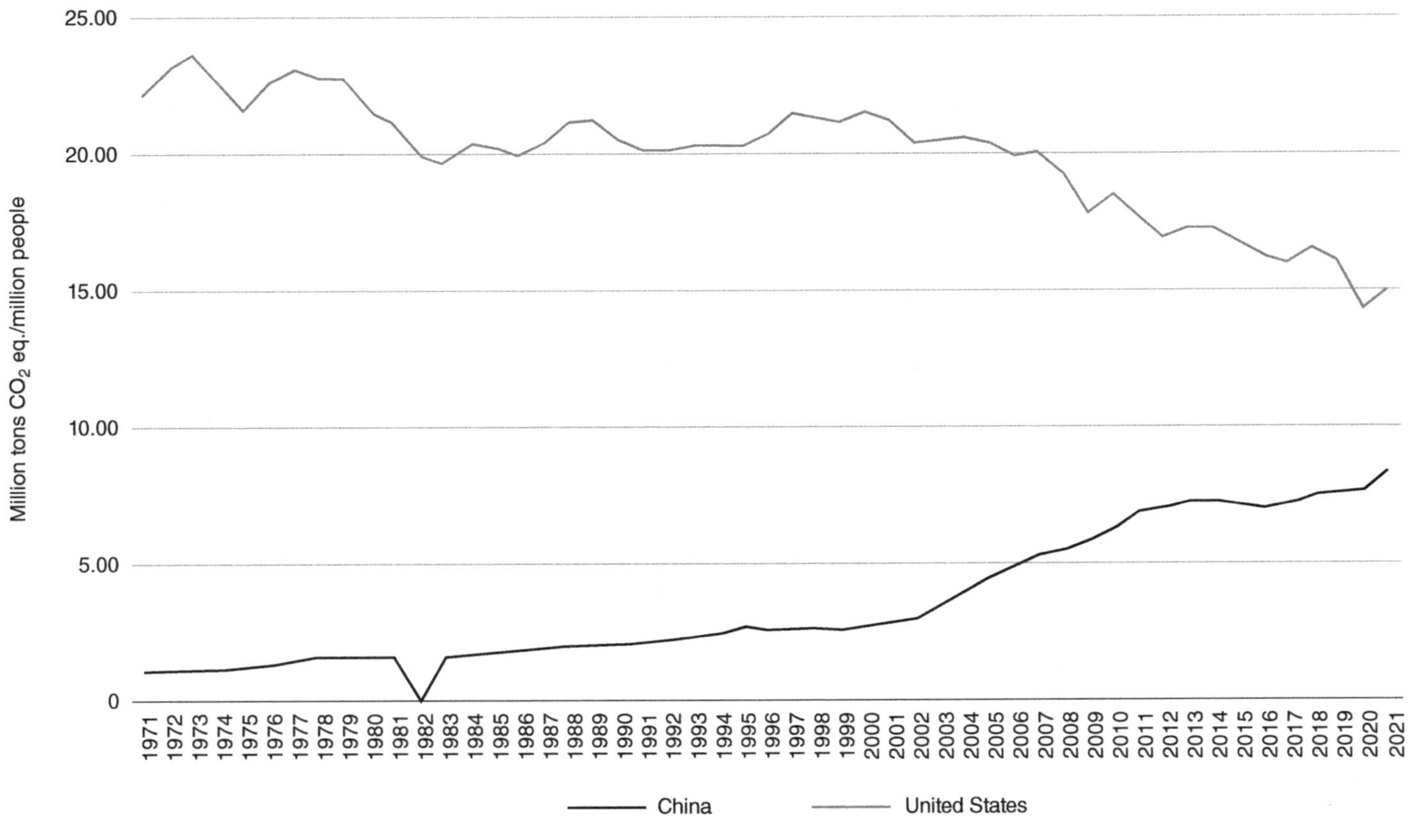

Figure 10.2 Per capita emissions: China versus United States.
Source: GHG emissions from energy from IEA, population data from World Bank.

hosting thousands of renewable energy projects funded by economically developed countries (Schroeder 2009). China's CDM experiences not only facilitated transfer of and foreign investment in clean technology but also made Beijing realize that clean energy development can be a "win-win" solution for economic development and climate mitigation (Qi and Wu 2013). Guided by this strategy, both the central and local governments created a range of policies since the mid 2000s to support homegrown industries of clean technologies, especially wind and solar power (Lewis 2013; Nahm 2017). With such government support, Chinese firms were able to establish unique capabilities for innovation in commercialization and scale up to mass production, so over the last decade the country began to lead the global production of clean energies (Helveston and Nahm 2019). Although coal still dominates China's energy mix, China is by far the global leader in terms of deployment of renewables, with its installed wind capacity accounting for 39 percent of the global share and solar capacity for 36 percent (Kyriakopoulou et al. 2022). In this respect, China seems to be making consistent progress in transitioning toward a low-carbon energy system through technological development and market scaling-up.

Moreover, as China's emissions continued to increase, the government increasingly began to set goals and targets on climate change. In 2011, response to climate change was for the first time mentioned as a chapter in China's Five-Year Plan – the most important policy document detailing the government's planned work. Since then, the government has continued to set specific targets on carbon intensity, energy intensity, and renewable energies in the subsequent 13th (2016–2020) and 14th (2021–2025) Five-Year Plans. The 14th Five-Year Plan also indicates the government's strong intention to align its development with the goals of emission peak before 2030 and carbon neutrality before 2060 by controlling both carbon intensity and total emissions. This trend of enhanced climate action in China can be partly attributed to the government's commitment to sustainable development due to its quest for performance legitimacy (Teng and Wang 2021). In the early 2010s, worsening environmental conditions caused public dissatisfaction across China and undermined the Chinese Communist Party's legitimacy (Fedorenko and Sun 2016; Wong and Karplus 2017). In response, China's top leaders introduced various environmental targets to measure government performance and proactively championed green development, even introducing the concept of "ecological civilization" as an overarching governance principle (Hansen et al. 2018).

With strong political motives and a rapidly expanding market on low-carbon technologies, the Chinese government was able to meet or even exceed almost all targets on climate change, especially those on renewable energy installed capacity (Lewis and Edwards 2021). Internationally, China has also felt increasing pressure to take more action to reduce its emissions after the Copenhagen conference in 2009, and since then has played a more prominent role in global climate governance and provided strong support for the Paris Agreement (Hilton and Kerr 2017). Taking such records into account and the recent momentum generated by China's carbon neutrality target, one can expect the country to become an important driver of global climate action.

However, the abovementioned progress is not without caveats. Coal remains dominant in China's energy mix and still accounts for more than half of China's primary energy consumption. Today, China's is still the world's largest consumer and producer of coal, so coal phaseout in China would significantly accelerate global decarbonization efforts. In 2020, Beijing announced the ambitious goals to peak carbon emissions by 2030 and achieve carbon neutrality by 2060. But in 2022 alone, China added 106 GW of new coal power capacity, the equivalent of two new plants per week (Myllyvirta et al. 2023). Partly due to insufficient effort to reduce fossil fuel use, China's total emissions continued to grow after 2016, and economic recovery from the first wave of the COVID-19 pandemic in 2021 even led to the country's largest emissions increase over the last ten years (Sandalow et al. 2022). In international fora, the Chinese government has also expressed its reluctance to rapidly phase out coal use in the country as demonstrated by the quest of China and India to use the wording of "phase down" instead of "phase out", when referring to unabated coal in the Glasgow Climate Pact (Depledge 2021). Therefore, many observers have criticized China's existing climate action as "highly insufficient" and demanded the Chinese government raise its ambition and specify how its planned emissions pathway can be made consistent with the Paris goals (Climate Action Tracker 2021). From this perspective, one can expect that the Chinese government's reluctance to strengthen or accelerate its climate mitigation efforts is due to resistance of some powerful incumbent actors who want to protect their interest in fossil fuels and coal in particular.

10.1.2 An Approach of State-Led Environmentalism

How can we understand China's policy response to climate change? This question requires an unpacking of the policymaking processes in the country and the interests of key actors. With an authoritarian political system, China's environmental and climate governance has been typically characterized by its state-driven processes. As pointed out by some experts, the relevant policymaking process is led by "consensus building" at the center in contrast to public and partisan debate in Western democracies, meaning the central government sets national targets and assigns responsibility of policy implementation to all levels of local governments and related companies (Qi and Wu 2013). In other words, this is a very top-down approach, which allows policymakers in Beijing to decide targets for the country with little involvement of non-state actors and the public. For this reason, some researchers label China's governance model as "authoritarian" environmentalism or "coercive" environmentalism (Gilley 2012; Li and Shapiro 2020). Indeed, research has shown that the state has always played a critical role in governing environmental change in China regardless of the specific instruments it may use (Guttman et al. 2021; Rooij et al. 2016; Sun 2022).

At the same time, simply interpreting China's climate governance as authoritarian environmentalism can be misleading. China's authoritarian system in the reform era has been described for a long time as fragmented due to the lack of coordination across government agencies and between central and local governments (Lieberthal and Lampton 1992; Mertha 2009). As a result, past research has shown much variation in local implementation of China's environmental and climate targets, which sometimes undermines the goals set by the national government (Alkon and Wong 2020; Eaton and Kostka 2014). As local governments may find leeway to interpret national targets and identify their own implementation strategy, businesses can accordingly lobby their local regulators to delay transition even if that was part of national policy. This might be especially the case for state-owned companies, which have a higher administrative status in the Chinese bureaucracy than provincial governments. Due to such fragmentation, many researchers remain pessimistic about China's climate ambition.

That said, there has been a recent, continuous trend of consolidating power to the center in China for environmental governance and green development, particularly since Xi Jinping took over in 2012 (Kostka and Zhang 2018). This trend has been likely driven in part by Xi's own interest in the environment and his stated intention to use ecological civilization to challenge the Western development model (Weins et al. 2023). More specifically, the central government has made a series of institutional arrangements, including through national campaigns and inspection systems, to ensure the achievement of targets by relevant subnational and non-state actors (Li and Shapiro 2020; Shen and Jiang 2021). In these ongoing processes, we have observed the increasing use of regulatory power by the central government to enforce change within local governments and businesses. As the Chinese state gained more influence over the market to push for green development, an optimistic view with respect to the politics of decarbonization is that, to continue and even accelerate transition, the government may have the capacity to weaken the power of incumbent actors who are likely to resist necessary changes.

In the next section, we use the case of power sector reform to show how, in China's unique political economy context, the state has tried to manage potential conflicts in the low-carbon energy transition.

10.2 Power Sector Reform and State-Owned Enterprise Restructuring

Restructuring state-owned power companies always been the central element of China's decade-long power sector reforms. Before 2002, China had a highly centralized power system, with only one state-owned enterprise (SOE) – State Power Corporation of China (or SPCC) – managing all power generation, transmission, and distribution assets across the country (Xu and Chen 2006). However, the Chinese government decided to terminate SPCC's monopolistic status by breaking it up into eleven companies in December 2002, during which power generation activities were separated from transmission and distribution services. In addition, the then existing power generation assets were redistributed among five newly created state-owned companies, which were encouraged to compete equally for developing new power generation infrastructures in China (Yeh and Lewis 2004). Meanwhile, the transmission and distribution system remained territorially monopolistic, with two utility companies responsible for northern and southern China respectively. The 2002 reform is often viewed as the beginning of a new "controlled but competitive" mode of power sector governance in China (Andrews-Speed and Zhang 2019; Baker et al. 2021).

The main purpose of the reforms in 2002 was to enhance the efficiency of SOEs, which was believed to be the main cause of a chronic power supply shortage since the start of China's marketization reforms in the 1980s. The effect was nearly immediate – power supply has been significantly enhanced since these reforms as illustrated in Figure 10.3, largely because of competition among the newly created "Big Five" power generation companies, known as Huaneng, Datang, Huadian, Guodian, and Zhongdiantou. The total installed capacity increased nearly fourfold between 2002 and 2016, from 356 GW to nearly 1650 GW. The "Big Five," together with several relatively smaller state-owned power generation companies (e.g. Guotou, Guohua, Huarun, Three Gorges, Zhongguanghe, and Zhongjieneng) have dominated the landscape of the power generation sector since then.[2] In addition, while China once welcomed foreign and private investments in the power sector (primarily in the late 1990s), most of these independent power producers were gradually squeezed out of the market due to competition from the leading SOEs (Baker et al. 2021). As a result, by the time China started to introduce the development of non-hydro renewables at scale beginning in the mid 2000s, its power sector was dominated by state-owned incumbents.

Renewable energy resources including wind and solar energy were never viewed as rivals to conventional energy resources by the Chinese government or SOEs in the earlier years of their development. At that time, conventional wisdom was that

[2] The Big Five plus China Three Gorges are also called the "Big Six" because the latter's capacity is closer to the Big Five.

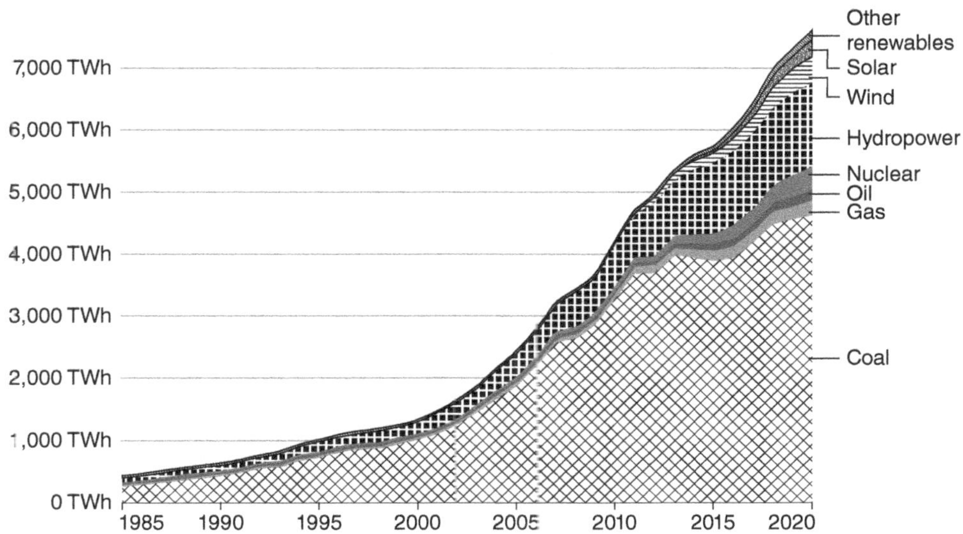

Figure 10.3 Power supply by source in China, 1985–2020.
Source: World Bank Open Data.

there was abundant space for both sectors to expand given a fast-growing economy and rising electricity demand (Shen and Xie 2018). In fact, wind and solar energy resources were viewed as tools to enhance China's energy security by increasing the country's self-reliance in electricity production. Since 2007, the introduction of feed-in tariffs and government procurement programs provided clear policy signals and notable incentives for some SOEs to venture into renewable energy investments.

Although not all leading SOEs were convinced of the promising prospect of renewable energy projects at that time, some believed that renewables would present a potential strategic opportunity. For example, Guodian developed impressive onshore wind energy capacities, mainly via its subsidiary company Longyuan Power, whereas Zhongdiantou became a leading developer of solar photovoltaic (PV) capacities. Even Shenhua Group, China's leading coal SOE, has developed several solar thermal projects around the country (Grikas 2016). While other SOEs lagged in renewable energy development during this initial period, many have caught up quickly in recent years.

10.2.1 Further Restructuring of the "Big Five" in the Era of Energy Transition

The "controlled and competitive model" largely resolved China's chronic power supply shortages, which was critical for China's fast-industrializing and fast-urbanizing economy. However, the fierce competition among leading

SOEs in the power generation sector ultimately led to a significant overcapacity problem. As economic growth slowed from more than 10 percent to less than 5 percent between the early 2010s and 2020, total energy consumption in China slowed. To avoid unnecessary competition among the SOEs and prevent massive "stranded" power generation capacity, the Chinese central government initiated another round of restructuring among SOEs to reallocate and coordinate the assets and resources. This restructuring started as early as 2011 when dozens of state-owned engineering, construction, and consultancy companies were merged into two leading infrastructure engineering, procurement, and construction (EPC) companies, known as PowerChina and China Energy Engineering Corporation (CEEC). In 2015, one of the Big Five, Zhongdiantou, was merged with the State Nuclear Power Technology Corporation to create the State Power Investment Corporation (SPIC), which makes it the only company in the Big Five with a comprehensive portfolio that includes coal and nuclear as well as wind, solar, and other clean energy assets. The restructuring reached its peak in 2017 when another Big Five company, the Guodian Group, was merged with Shenhua Group, the then largest coal-producing SOE in China. The merger deal created a mammoth organization – China Energy Investment Group (CHN Energy) – even compared to other SOEs, with total assets of more than USD 600 billion and 180 GW total installed capacity.

This restructuring significantly affected these SOEs' strategy for developing renewable energy capacities. The newly merged SPIC accelerated its investment in solar PV activities, and by 2019, it became the largest solar PV developer in the world in terms of accumulated capacity assets. It was also the first among the Big Five to achieve the 50 percent threshold of total installed generation capacity from renewable energy resources. Between 2015 and 2018, it was awarded twenty-one large-scale solar parks with a total of 2,175 MW under China's Solar PV Front Runner Program, a grand government procurement scheme aimed at expanding advanced and efficient technological solutions often in difficult contexts such as deserts or plateaus.

In China, the price of coal is largely determined by the markets, whereas electricity tariffs have been highly centrally regulated. As a result, turbulence in coal prices has significantly impacted the operations and economic viability of coal-fired power producers, which eventually led to mergers to reduce the structural uncertainties. CHN Energy was thus established as a result of the vertical merger between China's largest coal supplier, Shenhua, and the largest power producer, Guodian, which intended to mitigate the long-lasting conflicts between upstream and downstream activities in the thermal power sector. However, another notable impact of this merger is that it has created simultaneously the largest coal-fired and wind power producer in China, as the Longyuan Group under CHN Energy possesses more than 23.67 GW of wind energy capacity by 2021. The prospect of both coal-fired and wind power generation was reinforced as two pillars of the corporate strategy for CHN Energy. At the provincial level, a similar merger took place between two local SOEs in Shandong, Yankuang Group and Shandong Energy Group, which created a vertically integrated corporation in 2020 controlling both coal supply and power generation.

In 2023, another major SOE, China Three Gorges, reached a deal to hand over one of its major subsidiaries, China International Water & Electric Corporation (CWE), to the Zhongjiao Group. CWE has long been the leading dam builder around the globe, whereas Zhongjiao Group is the leading SOE for infrastructure and EPC construction, particularly in overseas markets. Such restructuring indicates that China Three Gorges will focus more on domestic markets in the future, particularly around the development and operation of hydro and solar energy capacities.

These cases illustrate the broader trend of major Chinese SOEs in the power generation sector being constantly restructured for different purposes over the past decade. Although the deployment of renewable energy capacity was not always the primary goal of these reforms, leading SOEs indeed consolidated their renewable energy capacities throughout these institutional changes. Notably, these reforms have simultaneously prevented the consolidation of fossil fuel interests in a few large power producers, instead incentivizing them to expand into renewable energy.

10.2.2 Transformed SOEs with Strong Interests in Clean Energy

Leading SOEs are indispensable forces in promoting the low-carbon transition in China, and ultimately helping the country achieve its dual goals of peaking carbon dioxide emissions before 2030 and achieving carbon neutrality before 2060. A key outcome of the Chinese central government's restructuring of these SOEs is that the total share of renewable energy capacity within these companies' portfolios increased substantially. By 2022, the share of renewable energy capacities within the Big Five's portfolios ranged between 31 percent and 66 percent and the annual additions to the wind and solar energy capacity reached 16 GW for SPIC, followed by Huaneng (13 GW), CHN Energy (11.8 GW), Huadian (7 GW), and Datang (4.5 GW) (see Table 10.1). The trend is expected to accelerate in the coming years, as the leading SOEs have been awarded many new projects that year, mainly through various procurement programs run by different provinces. It is estimated that among a total of 240 GW awarded wind and solar capacities in 2022, the Big Five plus China Three Gorges secured more than 100 GW. Accordingly, these SOEs have also announced ambitious medium-term targets for further expanding their wind and

Table 10.1 *Annual addition of renewable energy capacity in 2022 and targets by 2025 (in megawatts)*

	Annual addition in 2022 (MW)	**Targets 2021–2025 (MW)**	**Share of RE in total installed capacity (%)**
SPIC	16,000	50,000 (Solar)	65.87
CHN Energy	11,800	70,000–80,000	31
Huaneng	13,000	80,000+	41
Datang	4,500	50,000–80,000	42
Huadian	7,000	75,000	47.2
Three Gorges	N/A	70,000–80,000	n/a

Source: Sohu News and various online data sources.

solar energy capacities by 2025 (also see Table 10.1). In addition, all of them have established dedicated subsidiaries to compete in the growing renewables market for the expansion of wind and solar capacities. Over the past few years, some SOEs have already established distinctive advantages in particular technologies, such as SPIC in the solar PV sector and CHN Energy in the wind energy sector.

Moreover, onshore wind and solar PV projects have largely reached grid parity in China, which means that most of these newly contracted projects are no longer supported by state subsidies. As these wind and solar energy projects are becoming financially viable, they generate notable incomes that further justify their prominence within these SOEs. In comparison, the overall profitability of coal-fired power generation has been reduced significantly along with the curtailment of utilization hours. Although China's investment in coal-fired power generation capacity has increased recently due to concerns over energy security in the aftermath of power shortages in 2021, the actual coal-fired electricity outputs are expected to remain stable as many of these new capacities were added for providing ancillary services and are operating at very low-capacity factors.

A strong interest in renewables is increasingly being reflected in recent statements by relevant practitioners. For instance, an expert from a Big Five company admitted in 2022 that:

> nobody invests in coal-fired capacity for a profit nowadays. The money comes from the green energy sector, which is essential to keep the company going. It is just the opposite of the previous years when the green energy segment was heavily dependent on both government subsidies and revenues from coal-fired power capacities. The tables are now turned.[3]

[3] Interview conducted by an author in Beijing, July 2022.

Hence, after the government's reform to restructure these SOEs, leading Chinese power producers began to increasingly diversify their energy sources, and over time, entrench their interest in the clean energy market. In these processes, economic stability and energy security have always been a major concern of the Chinese government. At the same time, with strong commitments to climate action, policymakers in Beijing have been able to merge or regroup leading SOEs in the power sector, with the result being power sector incumbents did not form strong alliances to support fossil fuels and resist low-carbon transition.

10.3 Conclusions

The case of China's power sector reform shows that the state can strategically design policies to avoid potential conflicts with fossil fuel interests and promote the continuing effort for decarbonization. In other words, while it may be difficult to fully eliminate political conflicts between incumbents and rising actors, a strong state can use its regulatory power to manage such conflicts by reconciling different interests. That said, China has a unique political economy context where the state has authority over large SOEs, which are major players in the country's power market. The progress observed in China's power sector is also in the shadow of China's strong commitment to climate action and green development, namely the uptake of the norm to protect the planetary health by China's leadership and policymakers. Otherwise, the government would have had fewer incentives to continuously support renewable energy development.

To what extent can the insights gained from the China case shed light on comparative climate politics, including in other countries with very different political economy contexts? While acknowledging the importance of contextual factors, we suggest that when the state has the capacity to influence the market and regulate businesses, it can provide incumbents with incentives to take stronger action on climate change by reorienting the types of assets held by influential businesses (Colgan et al. 2021; Paterson 2021b). In recent years, many countries have begun to increasingly use industrial policies with the aim of accelerating decarbonization (Allan et al. 2021; Lewis 2021). But the government's support for new technologies alone often cannot address political conflicts between incumbents with entrenched interest in fossil fuels and rising green businesses. Accordingly, the case of Chinese power producer reforms sheds light on the importance of strategic management of political conflicts, and possible ways to prevent the formation or consolidation of anti-transition interests. To achieve such management, policymakers need to carefully design policies, including how to make and sequence institutional arrangements to shift the interest of powerful incumbents or weaken their resistance (Sewerin et al. 2022).

Lastly, we must add caveats to the Chinese case, as resistance to the low-carbon energy transition has certainly not been fully eliminated. To date, many practitioners in the country's power market may still be highly skeptical about the prospect of building a national energy system based on clean energy. Most recently, the power crisis in late 2021 led to the perception that an overly ambitious shift to clean energy was in fact the cause, as opposed to an overreliance on coal (Bloomberg News 2021). Accordingly, in order to continue and accelerate transition, the government needs to provide key actors with further incentives to reduce their interest in fossil fuels and increase their interest in clean energy. This would require the government to manage relevant conflicts in a durable way.

References

Alkon, M., and Wong, A. (2020). Authoritarian energy transitions undermined? Environmental governance cycles in China's power sector. *Energy Research & Social Science* 68, 101531. https://doi.org/10.1016/j.erss.2020.101531

Allan, B., Lewis, J. I., and Oatley, T. (2021). Green industrial policy and the global transformation of climate politics. *Global Environmental Politics* 21(4), 1–19. https://doi.org/10.1162/glep_a_00640

Andrews-Speed, P., and Zhang, S. (2019). *China as a Global Clean Energy Champion: Lifting the Veil* (1st ed.). Singapore: Palgrave Macmillan. https://doi.org/10.1007/978-981-13-3492-4

Baker, L., Shen, W., and Ayele, S. (2021). *Governing Procurement of Renewable Electricity Amid Power Sector Reforms*. The Institute of Development Studies and Partner Organisations. Report. https://opendocs.ids.ac.uk/opendocs/handle/20.500.12413/16976

Bernstein, S., and Hoffmann, M. (2019). Climate politics, metaphors and the fractal carbon trap. *Nature Climate Change* 9(12). https://doi.org/10.1038/s41558-019-0618-2

Bloomberg News. (2021). China's energy crisis has villagers questioning its climate path. *Bloomberg*, October 4. www.bloomberg.com/news/articles/2021-10-04/china-s-energy-crisis-has-villagers-questioning-its-climate-path

Climate Action Tracker. (2021). *Glasgow's 2030 Credibility Gap: Net Zero's Lip Service to Climate Action*. Climate Action Tracker Warming Projections Global Update Report.

Colgan, J. D., Green, J. F., and Hale, T. N. (2021). Asset revaluation and the existential politics of climate change. *International Organization* 75(2), 586–610. https://doi.org/10.1017/S0020818320000296

Depledge, J. (2021). Dr. Joanna Depledge: Reflections on Glasgow COP 26 beyond the headlines. Cambridge Zero, November 29. www.zero.cam.ac.uk/who-we-are/blog/dr-joanna-depledge-reflections-glasgow-cop-26-beyond-headlines

Eaton, S., and Kostka, G. (2014). Authoritarian environmentalism undermined? Local leaders' time horizons and environmental policy implementation in China. *The China Quarterly* 218, 359–380. https://doi.org/10.1017/S0305741014000356

Fedorenko, I., and Sun, Y. (2016). Microblogging-based civic participation on environment in China: A case study of the PM 2.c Campaign. *VOLUNTAS: International Journal of Voluntary and Nonprofit Organizations* 27(5), 2077–2105. https://doi.org/10.1007/s11266-015-9591-1

Gilley, B. (2012). Authoritarian environmentalism and China's response to climate change. *Environmental Politics* 21(2), 287–307. https://doi.org/10.1080/09644016.2012.651904

Grikas, M. (2016). Solarreserve and Shenhua group to partner in building 1,000 MW solar energy projects in China. Power Gen Advancement, May 3. www.powergenadvancement

.com/projects/solarreserve-and-shenhua-group-to-partner-in-building-1-000-mw-solar-energy-projects-in-china/

Guttman, D., Jing, Y., and Young, O. R., eds. (2021). *Non-state Actors in China and Global Environmental Governance*. Singapore: Palgrave Macmillan.

Hansen, M. H., Li, H., and Svarverud, R. (2018). Ecological civilization: Interpreting the Chinese past, projecting the global future. *Global Environmental Change* 53, 195–203. https://doi.org/10.1016/j.gloenvcha.2018.09.014

Helveston, J., and Nahm, J. (2019). China's key role in scaling low-carbon energy technologies. *Science* 366(6467), 794–796. https://doi.org/10.1126/science.aaz1014

Hilton, I., and Kerr, O. (2017). The Paris Agreement: China's "new normal" role in international climate negotiations. *Climate Policy* 17(1), 48–58. https://doi.org/10.1080/14693062.2016.1228521

Kostka, G., and Zhang, C. (2018). Tightening the grip: Environmental governance under Xi Jinping. *Environmental Politics* 27(5), 769–781. https://doi.org/10.1080/09644016.2018.1491116

Kyriakopoulou, D., Xia, L. Q., and Xie, C. (2022). Internationalism in climate action and China's role. Research article. Grantham Research Institute on Climate Change and the Environment, London School of Economics and Political Science. www.lse.ac.uk/granthaminstitute/wp-content/uploads/2022/01/Internationalism-in-climate-action-and-Chinas-role.pdf

Levin, K., Cashore, B., Bernstein, S., and Auld, G. (2012). Overcoming the tragedy of super wicked problems: Constraining our future selves to ameliorate global climate change. *Policy Sciences* 45(2), 123–152. https://doi.org/10.1007/s11077-012-9151-0

Lewis, J. I. (2013). *Green Innovation in China: China's Wind Power Industry and the Global Transition to a Low-Carbon Economy*. New York: Columbia University Press.

Lewis, J. I. (2021). Green industrial policy after Paris: Renewable energy policy measures and climate goals. *Global Environmental Politics* 21(4), 42–63. https://doi.org/10.1162/glep_a_00636

Lewis, J., and Edwards, L. (2021). Assessing China's energy and climate goals. Report. Center for American Progress, May 6. www.americanprogress.org/issues/security/reports/2021/05/06/499096/assessing-chinas-energy-climate-goals/

Li, Y., and Shapiro, J. (2020). *China Goes Green: Coercive Environmentalism for a Troubled Planet*. Cambridge: Polity.

Lieberthal, K., and Lampton, D. M., eds. (1992). *Bureaucracy, Politics, and Decision Making in Post-Mao China*. Berkely: University of California Press.

Mertha, A. (2009). "Fragmented authoritarianism 2.0": Political pluralization in the Chinese policy process. *The China Quarterly* 200, 995. https://doi.org/10.1017/S0305741009990592

Myllyvirta, L., Yu, A., Champenois, F., and Zhang, X. (2023). China permits two new coal power plants per week in 2022. Briefing, February. Center for Research on Energy and Clean Air. https://energyandcleanair.org/publication/china-permits-two-new-coal-power-plants-per-week-in-2022/

Nahm, J. (2017). Exploiting the implementation gap: Policy divergence and industrial upgrading in China's wind and solar sectors. *The China Quarterly* 231, 705–727. https://doi.org/10.1017/S030574101700090X

Paterson, M. (2021a). "The end of the fossil fuel age"? Discourse politics and climate change political economy. *New Political Economy* 26(6), 923–936. https://doi.org/10.1080/13563467.2020.1810218

Paterson, M. (2021b). Asset revaluation and beyond: Theorizing climate politics. E-International Relations, March 19. www.e-ir.info/2021/03/19/asset-revaluation-and-beyond-theorizing-climate-politics/

Qi, Y., and Wu, T. (2013). The politics of climate change in China. *WIREs Climate Change* 4(4), 301–313. https://doi.org/10.1002/wcc.221

Ritchie, H., Roser, M., and Rosado, P. (2020). CO_2 and greenhouse gas emissions. Our World in Data. https://ourworldindata.org/co2-and-greenhouse-gas-emissions

Roberts, C., Geels, F. W., Lockwood, M. et al. (2018). The politics of accelerating low-carbon transitions: Towards a new research agenda. *Energy Research & Social Science* 44, 304–311. https://doi.org/10.1016/j.erss.2018.06.001

Rooij, B. van, Stern, R. E., and Fürst, K. (2016). The authoritarian logic of regulatory pluralism: Understanding China's new environmental actors. *Regulation & Governance* 10(1), 3–13. https://doi.org/10.1111/rego.12074

Rosenbloom, D., Meadowcroft, J., and Cashore, B. (2019). Stability and climate policy? Harnessing insights on path dependence, policy feedback, and transition pathways. *Energy Research & Social Science* 50, 168–178. https://doi.org/10.1016/j.erss.2018.12.009

Sandalow, D., Meidan, M., Andrews-Speed, P. et al. (2022). *Guide to Chinese Climate Policy 2022*. Independently published.

Schroeder, M. (2009). Utilizing the clean development mechanism for the deployment of renewable energies in China. *Applied Energy* 86(2), 237–242. https://doi.org/10.1016/j.apenergy.2008.04.019

Sewerin, S., Cashore, B., and Howlett, M. (2022). New pathways to paradigm change in public policy: Combining insights from policy design, mix and feedback. *Policy & Politics* 50(3), 442–459. https://doi.org/10.1332/030557321X16528864819376

Shen, W., and Jiang, D. (2021). Making authoritarian environmentalism accountable? Understanding China's new reforms on environmental governance. *The Journal of Environment & Development* 30(1), 41–67. https://doi.org/10.1177/1070496520961136

Shen, W., and Xie, L. (2018). The political economy for low-carbon energy transition in China: Towards a new policy paradigm? *New Political Economy* 23(4), 407–421. https://doi.org/10.1080/13563467.2017.1371122

Stokes, L. C. (2020). *Short Circuiting Policy: Interest Groups and the Battle over Clean Energy and Climate Policy in the American States*. Oxford: Oxford University Press.

Sun, Y. (2022). *Certifying China: The Rise and Limits of Transnational Sustainability Governance in Emerging Economies*. Cambridge, MA: MIT Press.

Teng, F., and Wang, P. (2021). The evolution of climate governance in China: Drivers, features, and effectiveness. *Environmental Politics* 30(sup1), 141–161. https://doi.org/10.1080/09644016.2021.1985221

Weins, N. W., Zhu, A. L., Qian, J., Barbi Seleguim, F., and da Costa Ferreira, L. (2023). Ecological civilization in the making: The "construction" of China's climate-forestry nexus. *Environmental Sociology* 9(1), 6–19. https://doi.org/10.1080/23251042.2022.2124623

Wong, C., and Karplus, V. J. (2017). China's war on air pollution: Can existing governance structures support new ambitions? *The China Quarterly* 231, 662–684. https://doi.org/10.1017/S0305741017000947

Xu, S., and Chen, W. (2006). The reform of electricity power sector in the PR of China. *Energy Policy* 34(16), 2455–2465. https://doi.org/10.1016/j.enpol.2004.08.040

Yeh, E. T., and Lewis, J. I. (2004). State power and the logic of reform in China's electricity sector. *Pacific Affairs* 77(3), 437–465.

11

For Better or for Worse

A Break with Norway's Consensual Climate Tradition?

FAY M. FARSTAD, ERLEND A. T. HERMANSEN, AND BÅRD LAHN

Despite an extensive body of scholarship devoted to the task, significant disagreement remains about how to pursue low-carbon transitions. As outlined in Chapter 1, a key point of contention is about the merits of aiming for policy stability over time versus (re)politicization and more conflictual modes of challenging the status quo. At the heart of this debate lies different assumptions about the role of political and societal *consensus*, which may be seen either as a way of ensuring stability as policy lock-in and thus protecting climate policy measures against backlash or, alternatively, as a way of maintaining the status quo which therefore needs to be disrupted in order to promote climate action.

To feed into this debate, we focus on a country that is often seen as marked by a consensual political culture, and where climate policy development demonstrates key facets of both dynamics. Norway has for a long time aspired to be a leader in climate policy development, with a set of relatively stable policies and broad consensus regarding overall ambition (Boasson and Lahn 2017; Farstad 2019). Consensus has been underpinned by cross-party Climate Settlements (*Klimaforlik*) in 2008 and 2012, supported by all parties bar one (the right-wing populist Progress Party). Yet despite broad consensus about relatively ambitious climate goals, Norway has only reduced its domestic greenhouse gas (GHG) emissions by 4.7 percent between 1990 and 2022 (SSB 2023a). Moreover, in 2021, attempts to establish a new cross-party Climate Settlement similar to those in 2008 and 2012 failed. Does this failure to reach a new cross-party Climate Settlement constitute a break with Norway's consensual climate tradition, and is this good or bad news for climate policy?

In this chapter, we investigate whether and to what extent the consensus characterizing the 2000s and 2010s contributed to climate policy development or stasis, and why. We do this through analyzing two key sectors of Norwegian climate policy, namely petroleum policy and transport policy (more specifically electrical

This work was supported by the Research Council of Norway, grant number 268056.

vehicles – EVs). Constituting the two highest emitting sectors in Norway, these are critical cases to explore. Moreover, the two cases represent the largest sectors whose emissions are regulated by the European Union (EU) Emissions Trading Scheme (EU-ETS) and Effort Sharing Regulation (ESR) respectively, thus allowing us to assess the extent to which different regulatory systems affect (de/re)politicization. Through synthesizing documentary evidence, we analyze key developments within the two policy areas to explore the dynamic tensions between policy stability and (re)politicization.

The chapter is structured as follows. The first section outlines the key features of Norwegian climate policy, before the second and third sections outline the two cases – petroleum and EV policy respectively. For each sector, we attempt to identify reasons for de/repoliticization and its consequences. The final section discusses the findings and concludes.

11.1 Norwegian Climate Policy: Key Frameworks and Debates

Norway has been highlighted as an example of a corporatist state that actively includes new movements and issues into policy development, thus achieving a high degree of consensus across political divides as well as between state and civil society (Dryzek et al. 2003; cf. Grendstad et al. 2006). This dynamic is clearly identifiable in climate policy development. Norway was among the world's first countries to set a unilateral climate mitigation target – to stabilize emissions at 1989 levels by the year 2000 (Hovden and Lindseth 2004). A carbon tax was introduced in 1991 and an ETS in 2005 (which was incorporated into the EU-ETS in 2008). In 2007, the red–green coalition government under Prime Minister Jens Stoltenberg released a White Paper outlining a voluntary upgrade of Norway's original Kyoto Protocol target by 10 percent. The strategy also committed to an unconditional reduction of GHG emissions by 30 percent below 1990 levels within 2020, and carbon neutrality by 2050. After what has been characterized as a climate policy "bidding war," where the government also agreed to several major climate policy initiatives from the opposition, these ambitious goals were supported by all parties in the Norwegian parliament except the right-wing populist Progress Party (Hermansen 2015). The cross-party Climate Settlement reached in 2008 was reiterated and strengthened in 2012, following a similar logic of party competition in the negotiations. In the run-up to the Paris Agreement in 2015, Norway set itself the more stringent target of at least 40 percent emissions reductions by 2030 compared to 1990 levels, with the aim of being carbon-neutral by 2050. Norway ratified the Paris Agreement on June 21, 2016, and in 2022 adopted a revised target of a 55 percent reduction compared to 1990 levels by 2030. Climate policy is also underpinned by a Climate Act introduced in 2018.

Yet, despite a high level of consensus regarding relatively ambitious targets, Norway has only reduced its domestic GHG emissions by 4.7 percent between 1990 and 2022 (SSB 2023a). This is largely due to a booming petroleum sector and the fact that Norwegian electricity supply is essentially already decarbonized due to the prominence of hydropower. The country's emissions are therefore primarily related to petroleum, industry, transport, and agriculture, where emissions in many cases are costlier and technologically more difficult to abate, or require deeper structural changes to the economy. Norwegian climate policy has therefore been dominated by a logic of cost-efficiency, which has privileged international carbon trading over domestic emissions reductions as a means of achieving targets (Hermansen and Sundqvist 2022). The two major parties, Labour and Conservative, have historically functioned as veto players for setting stricter domestic mitigation targets, while smaller parties, such as the Christian Democrat, Liberal, and Socialist Left parties have traditionally argued for more ambitious domestic climate mitigation. Conflict over whether to meet climate commitments "at home" through domestic emissions reductions or "abroad" through international carbon trading has thus been a recurring topic in Norwegian climate policy (Hovden and Lindseth 2004), but agreement between Labour and the Conservatives ensured that the government largely relied on carbon trading to meet its commitments under the Kyoto Protocol (Hermansen and Sundqvist 2022). A high level of stability across shifting governments regarding overall policy ambitions (relatively ambitious international commitments) and the main policy instruments (carbon tax and trading) has thus coexisted with a relatively politicized debate marked by party competition on competing approaches to climate policy, which cut across the traditional left–right axis of Norwegian politics. This dynamic has had important implications for the link between climate and petroleum policy, as will be elaborated on in the next section.

In addition to already being part of the EU-ETS, in 2019 Norway chose to join the EU's ESR for 2030 (which covers emissions from sectors such as road transport, agriculture, buildings, and waste) and the Land Use, Land Use Change and Forestry (LULUCF) Regulation. The ESR sets an overall emissions reduction target for the ESR sectors in Norway but also includes the option to buy emission allocations from EU countries. The decision to join the ESR was supported by a majority in parliament. Some (e.g. Christensen 2018) argue that the decision to join the ESR is in line with the Norwegian tradition of keeping options open for emissions cuts abroad. Others (Farstad et al. 2022; Gulbrandsen and Hermansen 2022) suggest that the ESR decision "locks in" the need to cut emissions domestically through binding annual emission budgets.

The pressure to reduce domestic emissions has increased over time. In fact, the debate about *whether* to cut domestically has now largely been depoliticized as a consequence of adopting the ESR. However, with binding annual emissions budgets and an intensified pressure to cut emissions domestically by 2030, the ESR has also served to politicize climate policy, specifically about *how* Norway should reduce emissions. As Norway has already picked most of the "low hanging fruit" in terms of mitigation, governments are faced with increasingly difficult, costly, and controversial options for cutting GHGs.

Setting out how to reduce emissions in the ESR sectors, Prime Minister Erna Solberg's conservative coalition government published a climate White Paper in January 2021. However, despite initial statements that the government desired a new cross-party Climate Settlement on the plan similar to those in 2008 and 2012, it failed to establish one. This failure caused uproar and was labeled a "climate crash" (Nationen 2021). The government downplayed the lack of a new Climate Settlement, with Prime Minister Solberg arguing that there was no longer a need for such agreements. Whereas previous cross-party Climate Settlements had been vague deals about overall ambitions, the current plan was about concrete implementation, she argued. However, it is likely that the failure to establish a new settlement sprung from the fact that domestic climate policies were serving to bring existing lines of conflict and political cleavages to the fore. It also didn't help that a general election was coming up, with opposition parties keen to differentiate themselves from the government (Farstad and Aasen 2022).

The emphasis on domestic cuts continued under the minority coalition government consisting of the Labour and Centre parties elected in 2021. Their government declaration signaled they would meet the 2030 target solely through domestic action, which marks a sharp contrast to the previously dominant strategy of pursuing cuts through carbon trading. This turning point was brought about not only by Norway joining the ESR but also in response to meager domestic results historically and the 2021 election being hailed as a "climate election," thus ramping up party competition on the issue (Farstad and Aasen 2022).

Overall, we have therefore seen how Norway has developed relatively ambitious and stable climate policies and cross-party consensus over time, although this has become increasingly strained and politicized as the emphasis shifts to trickier domestic action. Are we witnessing a break with Norway's consensual climate tradition, and is this change for the better or for the worse? To answer these questions, we now turn to a more in-depth examination of the debates and developments within two key sectors, namely petroleum and EV policy.

11.2 Oil and Gas: Politicizing and Depoliticizing with Mixed Effects

Oil and gas extraction is Norway's most significant economic sector, at times accounting for half the country's export revenue. It is also the country's largest source of GHG emissions. It is therefore not surprising that the future of the oil industry has been one of the most contentious topics in climate policy discussions in recent years, and a dominant topic in the general election campaigns of 2017 and 2021 (Farstad and Aasen 2022). Seen in a longer perspective, however, this intense politicization is a rather recent phenomenon, as oil and gas extraction was generally not framed as a climate policy issue during the 1990s and 2000s (Bang and Lahn 2019). Environmental concerns over oil extraction tended to focus on the risk of oil spills and consequences for fisheries. Moreover, rather than focusing on the *production* of oil and gas for export, the 1990s and early 2000s saw intense debates about whether to increase the domestic *use* of natural gas through the construction of gas-fired power plants. In the remainder of this section, we first analyze the debates about gas power, which brought carbon capture and storage (CCS) technology to prominence. We then move on to the more recent politicization of the question of future oil and gas production.

11.2.1 Gas-Fired Power Plants and CCS

In 1995, the company Naturkraft applied for permits to construct three gas-fired power plants. The Labour and Conservative parties supported granting the permits, arguing that increasing emissions could be offset by buying emission credits from other countries. They were strongly opposed by smaller center and left parties, as well as environmental nongovernmental organizations (ENGOs), who argued that Norway should reduce emissions domestically and not introduce fossil fueled power to an effectively fully renewable energy system. The issue thus became a symbol of the party-political competition between the rival policy approaches of cutting emissions "at home" or "abroad," which came to dominate Norwegian political life well into the 2000s (Tjernshaugen 2011).

From a relatively early stage, the coalition opposing gas power plants demanded that CCS be a prerequisite for granting any permits. Members of this coalition had different motivations for supporting CCS: While some were enthusiastic about the technology, others saw requirements for an expensive and undeveloped technology as strategic leverage for obstructing permits (Tjernshaugen and Langhelle 2009). In 2000, the conflict came to a head when the centrist minority government under Prime Minister Kjell Magne Bondevik (Christian Democrats) refused to grant the permits. The parliamentary majority overruled the government, prompting the prime minister's resignation (Tjernshaugen 2011).

The new Labour government, headed by Prime Minister Jens Stoltenberg, immediately permitted the gas-fired power plants without requirements for CCS. Labour remained in power for only a year, however, and subsequent governments have all consisted of coalitions of parties supporting and opposing gas power. They have relied on different versions of a compromise in which existing permits are upheld but with a strong commitment to funding and encouraging CCS technology. When Bondevik returned as prime minister in a coalition with the Conservatives in 2001, the government increased R&D funding for CCS and established a state-owned innovation company to coordinate efforts (Boasson and Lahn 2017). Stoltenberg then made a comeback as prime minister in 2005, as leader of a red–green coalition between Labour, the Socialist Left, and Centre parties. His government allowed Statoil to construct a new gas-fired power plant at the Mongstad refinery, based on an agreement that the state and Statoil work jointly toward full-scale CCS within a few years after the power plant was built. The agreement, a compromise between the pro-gas Labour party and the anti-gas Socialist Left, was later branded by Stoltenberg as a major technology-development commitment akin to "Norway's moon landing" (Boasson 2015: 95).

Thus, the idea that CCS should be actively promoted won broad political support, albeit for different reasons: While the parties opposing gas power saw CCS requirements as a way of postponing new use of gas and preventing an increase in Norwegian emissions, the Labour and Conservative parties emphasized industrial and technological development. CCS thus came to function as a "political glue" that kept coalition partners with different climate policy views together (Tjernshaugen and Langhelle 2009) – or in other words, as a way of depoliticizing the controversial gas issue.

In the late 2000s, the appetite for gas-fired power waned. Only one of Naturkraft's power plants had been constructed, and it was later decommissioned due to high gas prices. The Mongstad power plant has also been slated for decommissioning, and the government has subsequently abandoned plans for full-scale CCS at existing power plants. A strong commitment to developing CCS technology remains, but the debate has shifted from bridging differences on the contentious gas issue toward a focus on technological and industrial development (Boasson 2015: 84). In 2020, the government launched the "Longship" project, which aims to establish infrastructure for transporting carbon from CCS projects to storage sites in the North Sea (Gassnova 2022). The government now focuses on CCS connected to industrial sites such as cement production and waste incineration, as well as potential "blue hydrogen" and direct air capture projects.

In summary, CCS has been used to both politicize and depoliticize the gas debate, with mixed effects. The politicization of gas-fired power plants in the 1990s likely prevented the construction of several power plants and placed climate policy at the top of the Norwegian political agenda. It also helped establish CCS as a priority within climate policy and led to increased funding for technology development. However, when CCS was primarily used as "political glue," it may have been less effective in terms of technological development, as several plans for full-scale CCS were later abandoned. In contrast, the more recent focus on industrial emissions and infrastructure for carbon transport and storage – a result of the gas power issue fading from the agenda and thus a depoliticization of the technology – could see a bigger long-term payoff in terms of developing CCS technology and new infrastructure for carbon storage.

At the same time, the politicization of gas and CCS may have had important spillover effects to other areas of climate and energy policy. When the Bondevik government resigned over the gas issue and the permits for gas power plants were granted in 2000, parliament also adopted a new objective of reducing energy demand and set up a state enterprise, ENOVA, to support energy efficiency in buildings. As Boasson (2015: 137) notes, this happened in a highly depoliticized manner, giving experts and administrators much leeway to develop policies. However, it clearly happened in a context of demand for "uncontroversial" climate policies that could bolster the government's green credentials following a polarizing fight over gas. This context likely created room for new initiatives in energy efficiency, cross-political support for new renewable energy production, and an increasing demand for ambitious climate policy in other areas.

11.2.2 Contesting the Future of Oil and Gas Production

While the domestic use of gas was highly politicized in the early 2000s, oil and gas extraction for export was less central to climate policy debates. This was partly a consequence of the international climate regime, which placed responsibility on the use rather than the production of fossil fuels (Bang and Lahn 2019). It was also partly made possible by Norway's embrace of carbon trading as a means of fulfilling its international commitments. Since the early 1990s, it had been clear that the emissions from oil and gas extraction would grow significantly in the coming decades, which would make it difficult for Norway to combine ambitious climate commitments with continued oil production. As a result, the Norwegian government became an active supporter of carbon trading in the negotiations leading up to the Kyoto Protocol, and subsequently relied on carbon credits to offset increasing emissions from its growing oil industry. This helped depoliticize oil and gas extraction as a climate policy issue, although there were still controversies regarding expanded oil exploration in the 2000s related to local environmental risks and negative effects on fisheries (Bang and Lahn 2019).

Since the 2013 parliamentary election, however, there has been a marked shift in the debate about oil production, with an increasing politicization of the future of the oil industry. The shift is closely linked to international developments. The formal adoption of a global temperature target provided a common yardstick for global climate action, which has been further translated into a budget for total allowable carbon emissions, and used by a range of actors to call for a halt to new oil and gas developments (Paterson 2021). In Norway, ENGOs shifted their focus from contesting oil exploration in specific environmentally sensitive areas toward demands for a full stop in exploration and a planned phaseout of Norwegian oil production (Bang and Lahn 2019). These demands were taken up by the smaller green parties in parliament – in particular the Green Party, which won its first parliamentary seat in the 2013 election by campaigning to end oil production (Farstad 2014). It also became a key demand in the large-scale school strikes organized across Norway in 2018–2019, and a heated issue during the 2021 general election (Farstad and Aasen 2022).

The response to the new challenges against further oil and gas extraction has been twofold. First, the larger parties have argued that any winding down of oil production will have to be market-led. The Labour and the Conservative parties have explicitly argued that climate change should be addressed on the demand-side rather than the supply-side, and that as long as there is a demand for fossil fuels, Norway should use its opportunities to meet that demand (Bang and Lahn 2019). Second, the oil industry has argued that Norway should aim to maintain or increase its share of global oil and gas markets because it is a more responsible, reliable, and democratic fossil fuel supplier with lower production-related emissions compared to Russia and OPEC. This line of argument has gained particular prominence following the 2022 energy crisis and cuts in Russian gas supplies to Europe.

However, the recent and intense politicization of oil production has so far led to few policy changes. A notable exception is a new commitment to assess the lifecycle emissions from oil production when approving new oil fields for development. In practice, however, such changes are unlikely to have much impact on the volume of Norwegian oil and gas extraction. In fact, policies for active exploration have been upheld, and oil production is likely to continue at high levels, fueled in part by COVID-19 support packages and increasing European demand for gas in light of the war in Ukraine.

11.3 A Generous EV Scheme Becomes Increasingly Effective, Expensive, and Politicized

The case of Norwegian EV policy has important parallels to the development of CCS policy, as it breaks with the prevalent discourse of cost-efficient cuts abroad. Moreover, part of the justification the political majority has provided for continuing exploration for new oil has been that climate policy should focus on managing the *demand* for fossil fuels, not the *production* (e.g. Helgesen 2016). Norway's ambitious EV policies can be seen as an example of consistency in this regard: While Norway will continue to produce oil to meet existing demand, the government's EV policy is designed to undercut the demand for its main export.

Norwegian EV policy has its roots in the early 1990s (Ryghaug and Skjølsvold 2019). A coalition of policy entrepreneurs comprising the ENGO Bellona, the pop group A-ha, and EV enthusiasts performed a number of public stunts, including civil disobedience (e.g. driving through toll roads without paying), which garnered public attention (Asphjell et al. 2013). Combined with prospects of developing an EV industry carried forward by technical entrepreneurs in the small EV company PIVCO (later Think), the interest organization NORSTART (later the Norwegian EV Association) and ENGOs, EVs were gradually given a number of advantages, such as no vehicle registration tax on purchasing new cars (and from 2001 no VAT), no annual vehicle license fee, free public parking, no fees on toll roads, and permission to drive in bus lanes (Ryghaug and Skjølsvold 2019).

Several of these policies, such as free public parking and passing through toll roads, had their origin at the local level, more specifically the municipality of Oslo (Ryghaug and Skjølsvold 2019). Over time, these advantages and subsidies became durable and stable policies at the national level, also because EV owners, although few initially, organized in the Norwegian EV Association, which has grown into a large and powerful organization campaigning to maintain advantages in order to phase in more EVs. Norway failed to develop a domestic EV industry, but when mainstream automakers such as Citroën, Mitsubishi, and Nissan started offering larger and safer EVs with a decent range, EVs became more widespread, particularly for commuters in suburbs (Figenbaum et al. 2015). However, it was not until Tesla (and later the large incumbent actors in the automotive industry) started to offer family cars with longer ranges that the diffusion of EVs really picked up. New policy initiatives, such as the cross-party Climate Settlements, National Transport Plans, and large-scale urban contractual agreements (Tønnesen et al. 2019) also contributed to facilitate EV diffusion. The result of these advantages, combined with green and cheap electricity, is that Norway has the highest number of EVs per capita in the world, making up 21 percent of the car fleet in 2022. The same year, 79 percent of new cars sold were EVs (SSB 2023b).

Most EV advantages have been backed by bipartisan support, as transport is a large emitting sector and there is increasing recognition about the need to cut emissions. A high level of car taxation in general, partly because Norway has no domestic car industry, has also allowed for powerful incentives through tax exemptions. However, the EV scheme is costly due to losses in tax income for the state, amounting to NOK 30 billion per year. Consequently, the EV subsidy scheme has been increasingly politicized, for example in the general election of 2021. Particularly leftist parties argued that "luxury subsidies" for expensive EVs (that also largely benefit people living in and around cities rather than rural areas) should be limited by, for example, introducing a fiscal ceiling. Consequently, in 2023 the Labour–Centre party coalition government reintroduced VAT on EVs for the part of the purchase that exceeds NOK 500,000. The smaller "climate parties," such as the Green, Socialist Left, and Liberal parties, disagreed with this move, arguing that it would slow the diffusion of EVs.

Thus, the initial success and subsequent politicization of EV policy can be explained by policy entrepreneurs, business prospects, and the absence of strong vested interests that eventually get overshadowed by discussions of social equality. Moreover, although Norway failed to develop a domestic EV production industry, the electrification of the transport sector has had a spillover effect on the shipping industry, where Norway tries to position itself in green shipping (Koasidis et al. 2020).

11.4 Discussion and Conclusions

There has been a constant and dialectic tension between de/repoliticization within petroleum and EV policy. The case of Norway thus provides important nuance to the debate about policy stability versus (re)politicization. It shows that a large degree of consensus on addressing climate change and stability around overall ambitions in terms of international commitments may coexist with intensely politicized debates regarding means, and that policy stability in some areas can go hand in hand with party-political competition over competing policy approaches in others.

On the one hand, depoliticization strategies in the case of EVs have ensured stability as policy lock-in, which has served to increase climate policy ambition. Likewise, the later depoliticization of the gas issue has allowed CCS technology to develop more rapidly, with an increasing focus on emission sources other than fossil fuels. On the other hand, the oil debate reveals depoliticizing strategies by the major political parties and vested interests, which have had the effect of ensuring stability in the sense of maintaining status quo for the oil and gas industry and making domestic emissions reductions more difficult.

The only overtly politicized policy development has been the early debate over gas and CCS, although this had mixed effects. While the debate might have prevented some gas-fired power plants from being developed, and created positive spillover effects for other areas such as energy efficiency in buildings, focusing on CCS connected to gas may not have been the best solution for developing the technology as a whole. Furthermore, the repoliticization of the oil debate has had negligible effects, perhaps underlining the significant power of vested fossil fuel interests. Likewise, the more recent politicization of EVs has served to weaken EV subsidies, with potential impacts on take-up and therefore emissions.

The spillover effects identified in the chapter are interesting, as they show that politicization of one policy area can spill over to another, such as in the case of

CCS and energy efficiency in buildings. Moreover, the CCS example demonstrates that despite the spillover effect resulting from a highly *politicized* debate, the affected policy area developed in a highly *de*politicized manner. Echoing the mixed effects highlighted in the chapter, this depoliticization allowed the area of energy efficiency in buildings to develop in a relatively ambitious manner as it allowed experts and administrators to "get on with the job." We see a similar spillover effect *within* sectors, such as in the case of EVs and green shipping, as well as the depoliticized EU-ETS making the ESR more politicized. We also reveal a multilevel spillover effect, whereby Europeanization and the decision to join the EU-ETS and ESR have affected national politicization, as well as local EV policy in Oslo affecting national-level policymaking. These spillover dynamics are significant and warrant further research in other cases and policy areas.

The variable success of the above de/repoliticization strategies can at least partly be explained by the factors influencing the de/repoliticization. Through our analysis, we have also identified explaining variables for de/repoliticization. On an overall level, *European integration* and Norway joining the ESR have served to both depoliticize and repoliticize climate policy through, on the one hand, "locking in" the need for domestic emissions reductions, while on the other hand creating heated debate about implementation. Likewise, over time, *party competition* has been a frequent explanation for repoliticization of climate policy, in most cases serving to increase ambition (e.g. relating to increasing domestic emissions targets). The exception is the case of the oil debate – whereas party competition has served to put the issue firmly on the political agenda, significant action has been hampered by the strength of *fossil fuel and vested interests*. Conversely, the absence of vested interests helps explain why EV policy has been easier to implement. The role of *entrepreneurs* also looms large in our review, with ENGOs acting as influential policy entrepreneurs in the case of petroleum policy, while the ENGO Bellona, EV enthusiasts, and local politicians were important technical entrepreneurs in the case of EV policy. Lastly, the *international climate regime* has influenced the dynamic tension between depoliticization and repoliticization, with the accounting rules under the Kyoto Protocol serving to depoliticize Norwegian petroleum expansion, while the Paris Agreement with its emphasis on the global carbon budget and domestic contributions has served to repoliticize the issue.

The mixed picture presented here thus reveals that depoliticization and (re)politicization can have both positive and negative effects on climate policy. However, it is important to emphasize that (re)politicization has rarely been about Norway's overarching climate goals or international commitments, with debates instead relating to *how* to achieve these. It is therefore worth noting that although we point to the occasional positive effects of (re)politicization, this is within a context of strong cross-party consensus and policy stability regarding an overall commitment to tackle climate change and take on relatively ambitious international obligations. (Re)politicization might therefore conceivably have different effects in different contexts, such as where there is more debate about overall targets and responsibility (such as in polarized countries like the United States). We therefore argue that it might be useful to differentiate between different "levels" of politicization – one at an overarching level relating to policy frameworks, goals, and targets, and one relating to implementation of specific policies.

Finally, we have seen how key Norwegian climate policies have developed through a dynamic tension of depoliticization and repoliticization. Through reviewing whether policy developments and debates have been chiefly depoliticizing or repoliticizing strategies and with what effects, we have revealed how – contrary to proponents of agonistic politics – depoliticizing strategies can indeed increase climate policy ambition in certain cases. However – contrary to proponents of policy stability and depoliticization – we have also seen how (re)politicization has served to ramp up ambitions in other cases. As such, we can surmise that not only is Norwegian climate policy less consensual than first assumed but the recent "crumbling of Norway's consensual tradition" might not be such bad news after all. Significantly, however, given the mixed effects revealed throughout, we argue that it is useful to embrace agnosticism in the debate over de/repoliticization, instead exploring this on an empirical and contextual basis.

References

Asphjell, A., Asphjell, Ø., and Kvisle, H. H. (2013). *Elbil på norsk*. Oslo: Transnova.

Bang, G., and Lahn, B. (2019). From oil as welfare to oil as risk? Norwegian petroleum resource governance and climate policy. *Climate Policy* 20(8), 997–1009.

Boasson, E. L. (2015). *National Climate Policy: A Multi-Field Approach*. Abingdon: Routledge.

Boasson, E. L., and Lahn, B. (2017). Norway: A dissonant cognitive leader? In R. K. W. Wurzel, J. Connely, and D. Liefferink, eds., *The European Union in International Climate Change Politics: Still Taking a Lead?* London: Routledge, pp. 189–205.

Christensen, L. (2018). Norges klimamål for 2030. Fra global til europeisk kostnadseffektivitet. *Norsk statsvitenskapelig tidsskrift* 34(2–03), 113–133.

Dryzek, J. S., Downes, D., Hernes, H. K., Hunold, C., and Schlosberg, D. (2003). *Green States and Social Movements*. New York: Oxford University Press.

Farstad, F. M. (2014). The Norwegian Greens: Coming in from the cold? *Environmental Politics* 23(6), 1096–1100.

Farstad, F. M. (2019). Does size matter? Comparing the party politics of climate change in Australia and Norway. *Environmental Politics* 28, 997–1016.

Farstad, F. M., and Aasen, M. (2022). Climate change doesn't win you a climate election: Party competition in the 2021 Norwegian general election. *Environmental Politics* 32(4), 732–742.

Farstad, F. M., Hermansen, E. A. T., van Oort, B., Grasbekk, B. S., and Brudevoll, K. (2022). Explaining radical policy change: Norwegian climate policy and the ban on cultivating peatlands. *Global Environmental Change* 74. https://doi.org/10.1016/j.gloenvcha.2022.102517

Figenbaum, E., Assum, T., and Kolbenstvedt, M. (2015). Electromobility in Norway: Experiences and opportunities. *Research in Transportation Economics* 50, 29–38.

Gassnova (2022). The Gassnova story. Website. https://gassnova.no/en/the-gassnova-story

Grendstad, G., Selle, P., Bortne, Ø., and Strømsnes, K. (2006). *Unique Environmentalism: A Comparative Perspective*. New York: Springer.

Gulbrandsen L. H. and Hermansen, E. A. T. (2022). Ever Closer Union? Norges tilknytning til EUs klimapolitikk [Norway's Climate Collaboration with the EU]. *Internasjonal Politikk* 80(1).

Helgesen, V. (2016). Tid for grønn omstilling. Verdens Gang, January 6.

Hermansen, E. A. T. (2015). Policy window entrepreneurship: The backstage of the world's largest REDD+ initiative. *Environmental Politics* 24(6), 932–950.

Hermansen, E. A. T., and Sundqvist, G. (2022). Top-down or bottom-up? Norwegian climate mitigation policy as a contested hybrid of policy approaches. *Climatic Change* 171, 26.

Hovden, E., and Lindseth, G. (2004). Discourses in Norwegian climate policy: National action or thinking globally? *Political Studies* 52(1), 63–81.

Koasidis, K., Karamaneas, A., Nikas, A. et al. (2020). Many miles to Paris: A sectoral innovation system analysis of the transport sector in Norway and Canada in light of the Paris Agreement. *Sustainability* 12(14), 5832.

Nationen (2021). Klimakrasj på Stortinget, March 25. www.nationen.no/politikk/klimakrasj-pa-stortinget/

Paterson, M. (2021). "The end of the fossil fuel age"? Discourse politics and climate change political economy. *New Political Economy* 26(6), 923–936.

Ryghaug, M., and Skjølsvold, T. M. (2019). Nurturing a regime shift toward electro-mobility in Norway. In M. Finger and M. Audouin, eds., *The Governance of Smart Transportation Systems* (e-book). Springer, pp. 147–165.

SSB (Statistics Norway) (2023a). Emissions to air. SSB, June 8. www.ssb.no/natur-og-miljo/forurensning-og-klima/statistikk/utslipp-til-luft

SSB (Statistics Norway) (2023b). Four out of five new cars in 2022 were electric. SSB, March 24. www.ssb.no/transport-og-reiseliv/landtransport/statistikk/bilparken/artikler/fire-av-fem-nye-biler-i-2022-var-elbiler

Tjernshaugen, A. (2011). The growth of political support for CO_2 capture and storage in Norway. *Environmental Politics* 20, 227–245.

Tjernshaugen, A., and Langhelle, O. (2009). Technology as political glue: CCS in Norway, in Meadowcroft, J., and Langhelle, O., eds., *Catching the Carbon*. Cheltenham: Edward Elgar, pp. 98–125.

Tønnesen, A., Krogstad, J., Christiansen, P., and Isaksson, K. (2019). National goals and tools to fulfil them: A study of opportunities and pitfalls in Norwegian metagovernance of urban mobility. *Transport Policy* 81, 35–44.

12

Stability and Politicization in Framework Climate Laws

DIARMUID TORNEY

Climate change has been characterized as a "super-wicked problem" (Lazarus 2009) requiring unprecedented and concerted responses across all sectors of economy and society. Since the enactment of the UK's pioneering Climate Change Act in 2008, framework climate laws have become an increasingly common governance response to the challenge of achieving the scale of transformation required. They are defined by Nash and Steurer (2019: 1053) as:

> framework legislation adopted by parliament that lays down general principles and obligations for climate change policymaking in a nation-state (or sub-state entity), with the explicit aim of reducing greenhouse gas (GHG) emissions in relevant sectors through specific measures to be implemented at a later stage.

Framework climate laws have been enacted across a growing range of countries in Europe and beyond. As of October 2023, twenty European countries have enacted such laws. In addition, the European Union's (EU) European Climate Law entered into force in July 2021.

The precise design characteristics of framework climate laws vary from country to country, but many share a range of core elements (Duwe and Evans 2020a). Most framework climate laws include a long-term quantitative target for GHG emissions reduction. Many also include intermediate GHG targets or mechanisms for setting such targets, provisions for government to set out policy measures to achieve those targets, and mechanisms for monitoring progress. Framework climate laws also typically provide some structured arrangements for incorporating scientific and expert advice into the policy process, usually through the establishment of a climate council, though their composition and functions vary considerably. They also to a greater or lesser extent provide opportunities for public participation in climate policymaking, though again this differs significantly from case to case (Duwe and Evans 2020b).

The standard characterization of framework climate laws is that they serve to provide stability in terms of climate policymaking (Averchenkova and Nachmany

2018; Duwe and Evans 2020a). This they do through a variety of institutional mechanisms, including enshrining GHG emissions targets in law and enhancing the technocratic quality of policymaking by strengthening the role of scientific advice in policymaking. The aim of this chapter is to provide a more nuanced assessment of the characteristics and roles of framework climate laws. I argue that, while some common design elements of framework climate laws do indeed serve to bring stability to climate policy, in many important respects, framework climate laws depart from the ideal design type envisioned by the literature on time inconsistency, commitment devices, and non-majoritarian institutions (NMIs). Moreover, in some ways framework climate laws can actually serve to make explicit the political conflicts at the heart of climate policy. By placing more or less binding GHG constraints, framework climate laws can make explicit trade-offs and political choices, including between economic sectors, and can thus serve to politicize even as they depoliticize. Moreover, by seeking to introduce stability to climate policymaking in the sense of *stability as policy lock-in*, framework climate laws simultaneously and deliberately seek to undermine and challenge *stability as the status quo*.

The chapter draws on examples of framework climate laws principally in European countries, which are used to illustrate the argument. The chapter does not aim to provide a comprehensive analysis. It relies on a combination of primary research and secondary sources, particularly the detailed comparative analysis conducted by Ecologic Institute on climate laws in Europe (Duwe and Evans 2020a). While the precise details of national climate laws differ, there are significant similarities between framework laws across jurisdictions. This is a somewhat surprising outcome, given the diversity of national circumstances, including political, legal, and institutional characteristics.

The chapter is structured as follows. The next section frames the discussion by providing a brief review of the literature on stability, delegation, and climate policy. The following four sections consider respectively four core design features typical of framework climate laws: enshrining targets in law, provisions for expert advice, arrangements for policy planning, and arrangements for monitoring and accountability. Across these elements, the chapter argues that, even as they aim to bring stability to climate policymaking, framework climate laws provide significant opportunities for the (re)politicization of climate policymaking.

12.1 Stability, Delegation, and Politicization in Climate Policy

Paterson, Tobin, and VanDeveer (2022) point to two contrasting strands in the literature on climate policy and politics, which emphasize respectively the need for policy stability and the need for politicization of climate policymaking. In Chapter 1, they elaborate on these themes and develop the themes of stability and (re) politicization in respect of climate policymaking. They distinguish between four forms of stability. Among these is *stability as policy lock-in*, which is a standard way that framework climate laws have been understood. This characterization builds on a longer strand of public policy literature focusing on the delegation of powers by governments to institutions that are not directly elected and not managed by elected politicians (Thatcher and Stone Sweet 2002). As a core component of these broader trends, so-called non-majoritarian institutions (NMIs) have proliferated in many policy areas and have taken different institutional forms, including independent regulatory agencies that are tasked with promoting competition and independent central banks.

Much of the literature on delegation adopts a rationalist principal–agent framework, focusing on the decision to delegate. According to this account, principals (governments) delegate to NMIs to resolve commitment problems, tie the hands of their successors, overcome information asymmetries in complex areas of governance, enhance the efficiency of rulemaking, and avoid taking blame for unpopular policies (Elgie and McMenamin 2005; Thatcher 2002; Wonka and Rittberger 2010). Among these drivers, the most prominent is the need for governments to be able to credibly commit to policies from which they have incentives to defect in the future, or the problem of time inconsistency. Kydland and Prescott (1977) were the first to identify the problem of *time inconsistency* in the context of economic policy. They proposed what they called "commitment devices" – "institutional arrangements that make it a difficult and time-consuming process to change the policy rules in all but emergency situations" – as a means of overcoming the problem of time inconsistency (Kydland and Prescott 1977: 487). In other words, commitment devices are mechanisms aimed at bringing stability as policy lock-in.

The problem of time inconsistency, and the need for policy stability, has been the focus of some attention in the literature on climate policy. Helm, Hepburn, and Mash (2003), for example, start from the observation that responding to climate change requires large-scale irreversible investments (in energy systems, for example), and that the profitability of those investments is very sensitive to climate policy decisions. They argue that a credible carbon policy must clear two hurdles, namely the need for government to define clear rules for resolution of trade-offs and to convince investors that it will not renege on their policy promises ex post. On the basis of this analysis, they advocate for the establishment of an independent "energy agency" analogous to the Monetary Policy Committee of the Bank of England.

In their discussion of the challenges of implementing long-term climate policy, Hovi, Sprinz, and Underdal (2009) also identify the problem of time inconsistency, as well as challenges of translating broad support for general measures into support for specific policy interventions and the problem of international cooperation in an anarchical international system. Brunner, Flachsland, and Marschinski (2012) highlight the trade-off between commitment and flexibility to adjust climate policy according to new information. They note in particular three areas of uncertainty in climate policy, namely the benefits of emissions abatement, the cost of emissions abatement, and international climate policy, pointing to the benefits of flexibility to update policy in light of new information.

Viewed from this perspective, framework climate laws can be viewed as mechanisms aimed at stabilizing climate policymaking, sending a clear signal to all sectors of economy and society about a collective and durable commitment to climate action (Duwe and Evans 2020b: 4). In the words of Averchenkova and Nachmany (2018: 110–111), enshrining climate commitments into law can limit the possibility of policymakers to "backtrack from earlier policy commitment … Embedding targets in law, as opposed to setting them informally through white papers or statements, makes them more difficult to change procedurally and politically." Similarly, Duwe and Evans (2020a: 12) write of framework climate laws that:

> establishing the system … in legal form makes it harder to go back on promises made. Laws can of course be changed, but legislation acts as a significant hurdle for policy rollbacks. A law is also a clear statement of sincerity. This sign of commitment … is also heard by external actors. Combined with a concrete long-term time horizon, this enhances certainty for all involved in the implementation.

As such, framework climate laws can be viewed as institutional mechanisms aimed at bringing stability as policy lock-in to climate policymaking. It is the contention of this chapter, however, that such a view misses some important characteristics of framework climate laws as they have been developed across a variety of jurisdictions, and that they in fact serve to politicize climate policy in important ways. In Chapter 1, Paterson, Tobin, and VanDeveer identify four forms of politicization. This chapter focuses in particular on *politicization as partisan competition* and *politicization as broader sociopolitical change*. In the following sections, I focus on four core design features typical of framework climate laws – enshrining targets in law, provisions for expert advice, arrangements for policy planning, and arrangements for monitoring and accountability.

12.2 Enshrining Climate Targets in Law

One common feature of framework climate laws is to enshrine climate change mitigation targets in law in order to introduce policy stability by indicating a clear direction of travel toward a decarbonized economy and society. These typically come in two varieties: medium-to-long-term targets (many around mid-century) and shorter-term targets (typically five to ten years). Most European framework climate laws have enshrined long-term quantitative targets for GHG emissions, and a majority have chosen net zero by 2050 as this goal (Duwe and Evans 2020b: 50–51). GHG targets can be expressed in a variety of ways (Hilson 2020). They can be binding or nonbinding. Most, but not all, European framework climate laws have enshrined a target for GHG emissions reduction in law (Duwe and Evans 2020b: 50–51). In Sweden, for example, GHG targets are not contained in the climate law itself but rather in a separate, nonlegislative "Climate Policy Framework for Sweden" (Nash and Steurer 2019; O'Gorman 2020). GHG targets can also be expressed as absolute or net emissions reductions. Many countries are aiming for "net zero" or some variation thereof, but national practice in regard to defining carbon sinks and reporting practices varies significantly (Hilson 2020).

The time frame of the long-term target has emerged as an important consideration. As Hilson (2020: 211) puts it, "[b]ecause, of course, the substantive target of net zero is itself already ambitious, the battleground over ambition has become focused on when it should be achieved." A majority of climate laws in Europe set net zero by 2050 as a central mitigation goal (Duwe and Evans 2020b: 50–51) but there are exceptions, with Sweden and Germany (the latter under its revised climate law) pledging net zero by 2045, and Finland by 2035. The majority of these targets, however, are in the medium-to-long-term, but they can have shorter-term effects by providing a benchmark against which nearer-term mitigation targets can be set and assessed. Only once we know where we want to get to by some date several decades hence is it possible to assess whether targets for 2025 or 2030, for example, are reasonable. In other words, a long-term target allows us to determine appropriate milestones on the way to that destination. That is not to say that a chosen pathway will necessarily involve a perfectly linear trajectory from today until the end point. There may be reasons, such as delayed carbon abatement of large-scale infrastructure investment, for limited backloading of emissions reduction over the coming decade, for example. But, absent a legally enshrined long-term goal, there is no benchmark against which intermediate targets can be assessed.

Intermediate targets are a second variety of target-setting in framework climate laws. These again often seek to bring stability to climate policymaking by "locking in" climate targets through legal means. The UK's Climate Change Act of 2008, the first of its kind worldwide that has been held up as a pioneer, took a distinctive approach (Carter 2014; Carter and Childs 2018). As well as setting quantitative GHG reduction targets for 2020 and 2050 (34 percent and 80 percent respectively, relative to 1990 levels),[1] the UK Climate Change Act also created a system of

[1] In June 2019, the Climate Change Act 2008 (2050 Target Amendment) Order 2019 amended the Climate Change Act to increase the 2050 emission reduction target from 80 percent to 100 percent, following advice from the Climate Change Committee.

legally binding five-year carbon budgets – overall limits on permissible GHG emissions for a five-year period. These carbon budgets are set by government 11.5 years in advance and can only be amended for very limited, defined reasons.[2] In a review of the UK Climate Change Act after ten years, Fankhauser and colleagues (2018) reported that the carbon budgets approach had transformed the UK power sector by providing a stable, long-term policy signal. However, they distinguished between stability over the overall target and stability in terms of particular policies. When making investment decisions, they noted, businesses and individuals look for certainty and stability about specific policy supports perhaps more than certainty about a broader direction of travel.

France has also adopted a carbon budget approach in climate legislation through the Energy Transition Act of 2015 and the Energy and Climate Act of 2019 (Duwe and Evans 2020a: 20). Germany has employed a variation on the budget approach, by breaking down a national emissions pathway to 2030 into annual values for the country as a whole and for each main sector of the economy and assigning responsibility for achieving these targets to relevant ministers with limited flexibility allowed toward meeting them. Ireland's 2021 Climate Action and Low Carbon Development (Amendment) Act, which significantly strengthened an earlier climate law passed in 2015 (Torney 2017), also introduced a system of carbon budgeting, modeled significantly on the UK Climate Change Act.

By enshrining GHG targets – either long-term or intermediate – in primary legislation, framework climate laws seem to operate as classic commitment devices, aiming to introduce policy stability over time. Primary legislation is procedurally more difficult and politically challenging to row back on compared with nonstatutory targets. However, even this most emblematic of stabilizing instruments contains within it the seeds of repoliticization. Enshrining targets in law opens up the possibility of legal remedy through the courts. There has been an increasing trend toward strategic climate litigation since 2015 along with key judgments by apex courts (Setzer and Higham 2021; UNEP 2023). Indeed, White and O Callaghan-White identify "a strong interventionist trend in the approach of domestic courts in Europe to the issue of climate change" and suggest that a "domino effect" in climate litigation that, they argue, will "heighten the sensitivity of policymakers and significantly increase the evaluation of government action (and inaction) on climate change" (White and O Callaghan-White 2021: 2). Recent high-profile successful climate litigation cases, including in the Netherlands, Ireland, and Germany, have focused on the adequacy of climate targets or the adequacy of government policy plans.

[2] These circumstances are: (i) changes in scientific knowledge about climate change; (ii) changes European or international law or policy; or (iii) changes in the scope of greenhouse gases included within the target or the treatment of emissions from aviation and shipping in national targets (UK CCA, S.6)

Pursuing legal action is an important avenue for citizens and environmental groups to challenge the adequacy of government policy responses to climate change. It is striking that a number of strategic litigation cases brought by citizen groups have included a significant element of popular mobilization, including the Urgenda case in the Netherlands and the case brought by Friends of the Irish Environment in Ireland (commonly known as Climate Case Ireland). Nonetheless, it is also arguably the case that a focus on legal remedy in the event of failure to meet GHG emissions targets is somewhat misplaced. As Reid (2012) argues in the case of the UK Climate Change Act, accountability of the government to parliament rather than through the courts is the primary accountability mechanism within that law.

12.3 Arrangements for Incorporation of Expert Advice

Another central element of many framework climate laws is the creation of expert bodies to input into climate change policymaking. Averchenkova and colleagues (2018: 2) argue that independent expert bodies can "strengthen climate governance by introducing a long-term perspective, enhancing the credibility of climate targets and ensuring more evidence-based policymaking." However, in contrast with other policy areas, such as monetary policy in which important policymaking functions are delegated to independent technical experts to insulate decisions from political control, independent expert bodies in climate policymaking are generally imbued with advisory rather than policy-setting powers. Moreover, not all instances of delegating climate policy to NMIs result in more ambitious climate policy. Burns and Tobin (2020) find that delegated and implementing acts have been used to dismantle EU climate policy.

In national framework climate laws, independent scientific climate councils typically serve to play roles as watchdogs, advisors, and conveners (Evans and Duwe 2021). The *watchdog* role involves producing regular independent assessments of government action or inaction. The strength of this role is conditioned by the body's available resources and also whether the government is required to formally respond to its recommendations. In Denmark, France, and the UK, for example, the government is required to respond to recommendations (Evans and Duwe 2021: 36). The *advisor* role entails provision of input to climate policy formation. In the UK, the Climate Change Committee plays an important role in providing advice on the setting of carbon budgets, which is enhanced by the fact that the Climate Change Committee makes the first recommendation to which the government must respond rather than the other way around. The Danish climate council is required to prepare a catalogue of possible climate policy instruments for consideration by the government. The third role is as a *convener*. Some climate councils engage in stakeholder outreach to varying extents. The most developed of these is the Danish climate council, which is tasked with managing a dedicated public and stakeholder dialogue mechanism.

In terms of the composition of advisory bodies, Evans and Duwe (2021) distinguish between those composed essentially only of scientists and other academic experts and those composed of a wider range of stakeholders. In general, independent scientific climate councils are typically composed of external experts who do not represent particular interest groups or government departments or bodies (Duwe and Evans 2020b; Evans and Duwe 2021), though there are some exceptions such as Ireland, where the principals of some state bodies serve as *ex officio* members (Weaver, Lötjönen, and Ollikainen 2019). Such composition increases the perceived independence and objectivity of climate councils, and also enhances their role as "knowledge-brokers" by building bridges between scientific research and policymaking (Evans and Duwe 2021; Weaver, Lötjönen, and Ollikainen 2019). Some climate laws, such as the UK Climate Change Act, stipulate that the climate council should include a diversity of expertise. In a number of cases, including in Denmark, Finland, France, Sweden, and Switzerland, the councils themselves self-select their members, who are then officially appointed by government with varying degrees of governmental oversight (Evans and Duwe 2021: 34).

While climate advisory bodies play a variety of arguably important roles, all of this is some distance away from an ideal-type technocracy in which policymaking functions are delegated to independent experts with a view to bringing stability to climate policy. Indeed, Helm, Hepburn, and Mash (2003) considered a model in which an independent agency would be given policy autonomy to pursue any necessary measures to achieve a specified GHG emissions reduction goal and explicitly rejected it on the basis that no government would ever agree to such delegation. So it has proved to be in practice.

12.4 Arrangements for Policy Planning

As well as setting targets for various time horizons or provisions for how such targets must be set, framework climate laws also set out various requirements on governments to develop policy plans to meet those targets, incorporating both longer-term strategies and shorter-term implementation plans. Only some framework climate laws, including those of Finland, France, and Spain, set specific arrangements for longer-term strategic climate policy planning. Many more framework climate laws provide arrangements for more detailed climate policy development over shorter time frames, usually a ten-to-fifteen-year time horizon (Duwe and Evans 2020b). These processes are typically repeated on a regular cycle of four to five years, with the Finnish and Swedish framework climate laws linking these to regular electoral cycles (Nash and Steurer 2019). Some framework climate laws, such as Denmark's, also provide for annual climate policy planning in their framework climate laws. It should also be noted that the EU governance framework for climate action (EU Regulation 2018/1999) requires member states to develop long-term climate strategies with a thirty-year time horizon and national energy and climate plans with a ten-year time horizon, both to be updated every five years.

Some framework climate laws also make provision for adaptation policy planning, including requirements to produce periodic climate risk assessments and planning frameworks for adaptation to climate change. The UK Climate Change Act requires the government to produce five-year risk assessments and adaptation plans, as well as providing for independent evaluation of risk assessments and plans. The French climate law requires asset owners and managers to report on climate change risks (World Bank 2020). The new European Climate Law requires all member states to adopt, implement, and regularly update national adaptation strategies and plans (Article 5).

Generally, the requirements set out in framework climate laws for policy planning are procedural in nature – stipulating in what way and over what time periods policymaking must occur – and do not extend to prescribing specific policy instruments. There are, however, some exceptions. The French climate law, for example, includes a range of specific policies, including amendments to existing legal codes covering carbon tax, CO_2 performance standards for thermal power plants, and renovation obligations. The Spanish climate law includes a ban on the sale of combustion engine vehicles by 2040 (Duwe and Evans 2020b: 25). Some framework climate laws also make provision for climate mainstreaming and connecting the climate policy cycle with the annual budget process, including in France, Germany, and Sweden (Duwe and Evans 2020b: 26; World Bank 2020). The French climate law makes provision for mainstreaming climate action into all government policymaking. The French climate law also places climate reporting obligations on financial institutions. Such enshrining of sectoral targets and policies in primary legislation – as distinct from overall targets for GHG emissions reductions – act more strongly to bring stability to the policy landscape. For the most part, however, arrangements for policy planning in framework climate laws impose procedural obligations on governments. These procedural obligations are, nonetheless, anchored by the legally enshrined targets that underpin them. That is to say, the policy plans and instruments developed under the procedural requirements set out in framework climate laws are required to be consistent with the GHG reduction targets set out in law or put in place under the provisions of the law (in the case of carbon budgets and equivalent).

While such measures aim to bring stability in terms of policy design over time, they can in fact serve to destabilize the status quo, bringing to the fore partisan political conflict and inter-sectoral trade-offs. In Ireland, for example, the process of setting so-called sectoral emissions ceilings – the division of overall statutory carbon budgets into sectoral shares, which each sector would be responsible for delivering – generated hugely contentious political conflict over the relative share of the total emissions reduction effort that would be taken on by the agricultural sector, which is one of the most powerful interest groups in the Irish economy and society (Torney and O'Gorman 2019).

12.5 Monitoring and Accountability

Progress monitoring is a core component of framework climate laws, but whether these arrangements serve to bring stability or to repoliticize climate policy depends on the institutional form and requirements on government to respond. Typically, framework climate laws provide a basis for evaluation and feedback through some combination of government reporting along with evaluation from more or less independent advisory bodies (Nash and Steurer 2019). Key questions in this regard concern which institutions are given responsibility for monitoring and reporting, and what mechanisms are in place to require a response to monitoring reports. Different national climate laws vary in terms of which government entity is responsible for reporting. In Sweden, Germany, and Denmark, responsibility for annual progress reporting lies with government, whereas in France and the Netherlands it is the scientific expert body that reports. In the UK, both the government and the Climate Change Committee are required to issue annual progress reports (Duwe and Evans 2020b: 27).

Framework climate laws also vary in terms of what obligations the government is under to respond to progress reviews, and particularly whether and in what ways progress monitoring can trigger requirements for additional action to address shortfalls. However, Duwe and Evans (2020b) caution that such provisions can become "a formality without consequences" unless there is a clear and defined sequence of steps that must be undertaken in such circumstances. The German Climate Protection Law provides perhaps the most elaborate version of such an arrangement. The country's 2030 target introduced national emissions pathways for each sector of the economy, with responsibility for meeting sectoral targets assigned to the ministry most responsible for that sector. The federal government is required to report annually on GHG emissions in each sector. If this reporting shows that emissions for the preceding year have exceeded the sectoral emissions limits set out in the law, the responsible government ministry is required, within three months, to present an "immediate action program for the relevant sector" that "shall ensure compliance with the annual sectoral emission budgets in the subsequent years" (Federal Ministry of Justice 2019).

In the UK, the government is formally required to respond to the annual reports of the Climate Change Committee. In addition, in the event of a carbon budget not being met, the secretary of state must explain to parliament why the budget has not been met and set out proposals to compensate for excess emissions in future periods. In France, the government is required to respond within six months to the scientific advisory body's annual report, and to indicate how any gaps shortfalls in emissions reduction will be addressed through additional actions. In the Netherlands, a two-yearly progress report on the climate plan can trigger additional policy actions if required. In Denmark, the government is required to consider the need for additional measures when developing its annual climate program (Duwe and Evans 2020b: 25). Under Ireland's revised climate law, the Climate Change Advisory Council is mandated to publish an annual review of government progress, and in turn the government can be required to account to a parliamentary committee regarding the findings of the Advisory Council's annual review (Torney 2021).

There is, in short, a diversity of approaches to monitoring and accountability in framework laws. This diversity includes variation in terms of the degree to which different branches of government – parliament and executive – are assigned responsibility for ensuring that commitments are adhered to. This variation, in turn, is likely to shape the degree to which accountability mechanisms politicize climate policy by, for example, bringing it explicitly into the domain of parliamentary politics.

12.6 Conclusions

Framework climate laws have typically been viewed in the literature and by proponents as legal and institutional mechanisms to introduce greater stability into climate change policymaking. However, upon closer examination the picture is more nuanced in at least three ways. First, despite the appearance of depoliticization and technocracy, framework climate laws do not remove politics from climate change policymaking. Enshrining targets in law may hide politics but it does not remove it. Moreover, the functions of technocratic expert bodies in climate policy are quite different from how NMIs have been designed in other policy spheres. In a range of policy arenas, such institutions are granted executive decision-making powers, such as in the case of central banks and regulatory institutions. By contrast, climate change expert advisory bodies as established under framework climate laws are exactly that – advisory rather than executive in nature.

Second, by providing one form of stability, framework climate laws actively diminish another. Returning to the distinction made in Chapter 1, while framework climate laws seek to introduce stability to climate policymaking in the sense of *stability as policy lock-in*, they simultaneously and deliberately seek to undermine and challenge *stability as the status quo*. This perhaps unavoidably generates political conflict, mobilizing status quo actors to defend their interests through a variety of means and serving to (re)politicize climate policymaking by bringing difficult choices such as sectoral trade-offs to the forefront of the political arena.

A third way in which framework climate laws serve to (re)politicize climate policymaking is by providing new institutional openings for political mobilization. The degree to which these institutional openings are created depends to some extent on the institutional design of specific climate laws, but two broad trends can be observed. The first stems from the legalization of climate targets, which provides new routes for mobilization through judicial challenge. We have seen such mobilization occur in a number of different settings, with climate litigation becoming an increasingly prominent phenomenon, including challenges brought under framework climate laws as can be seen, for example, in Germany and Ireland. The second new opening for politicization through framework climate laws comes through the mechanisms of parliamentary accountability they introduce. By requiring governments to submit to parliamentary scrutiny, framework climate laws bring climate policymaking firmly into the partisan political sphere.

Overall, what emerges from this analysis is a complex picture in which framework climate laws serve both to stabilize and to destabilize climate policymaking and provide new openings for politicization in the spheres of parliamentary politics as well as civil society mobilization. These conclusions support the contention of the volume as a whole that the relationship between stability and politicization should be seen as much more complex than a simple binary view of the two concepts would suggest.

References

Averchenkova, A., and Nachmany, M. (2018). Institutional aspects of climate legislation. In A. Averchenkova, S. Fankhauser, and M. Nachmany, eds., *Trends in Climate Change Legislation*. Cheltenham: Edward Elgar, pp. 108–122.

Brunner, S., Flachsland, C., and Marschinski, R. (2012). Credible commitment in carbon policy. *Climate Policy* 12(2), 255–271.

Burns, C., and Tobin, P. (2020). Crisis, climate change and comitology: Policy dismantling via the backdoor? *JCMS: Journal of Common Market Studies* 58(3), 527–544.

Carter, N. (2014). The politics of climate change in the UK. *WIREs Climate Change* 5, 423–433.

Carter, N., and Childs, M. (2018). Friends of the Earth as a policy entrepreneur: "The Big Ask" campaign for a UK climate change act. *Environmental Politics* 27(6), 994–1013.

Duwe, M., and Evans. N. (2020a). *Climate Laws in Europe: Good Practices in Net-Zero Management*. Berlin: Ecologic Institute.

Duwe, M., and Evans. N. (2020b). *Professionalizing Climate Policy via Legislation: Good Practices in Long-Term National Climate Framework Laws*. Tel Aviv: Israel Public Policy Institute and Heinrich Böll Foundation Tel Aviv.

Elgie, R., and McMenamin, I. (2005). Credible commitment, political uncertainty or policy complexity? Explaining variations in the independence of non-majoritarian institutions in France. *British Journal of Political Science* 35(3), 531–548.

Evans, N., and Duwe, M. (2021). *Climate Governance Systems in Europe: The Role of National Advisory Bodies*. Berlin: Ecologic Institute & IDDRI.

Fankhauser, S., Averchenkova, A., and Finnegan, J. (2018). *10 Years of the UK Climate Change Act*. London: Grantham Research Institute on Climate Change and the Environment.

Federal Ministry of Justice. (2019). Federal Climate Change Act (Bundes-Klimaschutzgesetz). www.gesetze-im-internet.de/englisch_ksg/englisch_ksg.html.

Helm, D., Hepburn, C., and Mash, R. (2003). Credible carbon policy. *Oxford Review of Economic Policy* 19(3), 438–450.

Hilson, C. (2020). Hitting the target? analysing the use of targets in climate law. *Journal of Environmental Law* 32, 195–220.

Hovi, J., Sprinz, D. F., and Underdal, A. (2009). Implementing long-term climate policy: Time inconsistency, domestic politics, international anarchy. *Global Environmental Politics* 9(3), 20–39.

Kydland, F. E., and Prescott, E. C. (1977). Rules rather than discretion: The inconsistency of optimal plans. *Journal of Political Economy* 85(3), 473–492.

Lazarus, R. J. (2009). Super wicked problems and climate change: Restraining the present to liberate the future. *Cornell Law Review* 94, 1153–1233.

Nash, S. L., and Steurer, R. (2019). Taking stock of climate change acts in Europe: Living policy processes or symbolic gestures? *Climate Policy* 19(8), 1052–1065.

O'Gorman, R. (2020). Climate law in Ireland: EU and national dimensions. In D. Robbins, D. Torney, and P. Brereton, eds., *Ireland and the Climate Crisis*. London: Palgrave, pp. 73–89.

Paterson, M., Tobin, P., and VanDeveer, S. D. (2022). Climate governance antagonisms: Policy stability and repoliticization. *Global Environmental Politics* 22(2), 1–11.

Reid, C. T. (2012). A new sort of duty? The significance of "outcome" duties in the Climate change and child poverty acts. *Public Law* 4, 749–767.

Setzer, J., and Higham, C. (2021). *Global Trends in Climate Change Litigation: 2021 Snapshot*. London: Grantham Research Institute on Climate Change.

Thatcher, M. (2002). Delegation to independent regulatory agencies: Pressures, functions and contextual mediation. *West European Politics* 25(1), 125–147.

Thatcher, M., and Sweet, A. S. (2002). Theory and practice of delegation to non-majoritarian institutions. *West European Politics* 25(1), 1–22.

Torney, D. (2017). If at first you don't succeed: The development of climate change legislation in Ireland. *Irish Political Studies* 32(2), 247–267.

Torney, D. (2021). *Comparative Assessment of the Climate Action and Low Carbon Development (Amendment) Act 2021*. Dublin: Dublin City University. https://diarmuidtorneyorg.files.wordpress.com/2021/10/climate-law-assessment-final-7-october.pdf

Torney, D., and O'Gorman, R. (2019). A laggard in good times and bad? The limited impact of EU membership on Ireland's climate change and environmental policy. *Irish Political Studies* 34(4), 575–594.

UNEP (United Nations Environment Programme). (2023). *Global Climate Litigation Report: 2023 Status Review*. Nairobi: UNEP. https://wedocs.unep.org/bitstream/handle/20.500.11822/43008/global_climate_litigation_report_2023.pdf?sequence=3

Weaver, S., Lötjönen, S., and Ollikainen, M. (2019). *Overview of National Climate Change Advisory Councils*. Helsinki: Finnish Climate Change Panel.

White, A., and O Callaghan-White, L. (2021). *Taking Governments to Court Climate: Litigation and Its Consequences*. Dublin: Institute of International and European Affairs.

Wonka, A., and Rittberger, B. (2010). Credibility, complexity and uncertainty: Explaining the institutional independence of 29 EU agencies. *West European Politics* 33(4), 730–752.

World Bank. (2020). *World Bank Reference Guide to Climate Change Framework Legislation*. Washington, DC: World Bank Group.

13

The Illusion of "Apolitical" Climate Policy

MATTO MILDENBERGER AND MATTHEW LOCKWOOD

A widespread belief exists that effective climate policymaking requires the establishment of independent institutions to insulate decision-making from short-term political pressures and provide stability for investors and other actors (Knaggård and Pihl 2015; Dudley et al. 2021). From this perspective, the time structure of the climate change problem makes climate policy similar to an investment; the benefits are mainly in the distant future, while costs must be incurred up front (Jacobs 2011; Lazarus 2009). As a result, while governments might make initial climate policy commitments, they may also face incentives for policy retrenchment or reversal if short-term policy costs generate public backlash. Policies may also be reversed when governing parties change. Because of this "time inconsistency" problem, a government's stated commitments may not be credible, especially to investors, who must be convinced that either current or future governments will not reverse the policy and strand their investments (e.g. Bosetti and Victor 2011). The belief is that well-designed institutions to remove climate policy decision-making from the realm of party politics can solve this central policymaking challenge and provide policy stability (see Paterson, Tobin, and VanDeveer, Chapter 1, this volume).

This analysis and the concept of time inconsistency was originally developed in monetary policy by Kydland and Prescott (1977), where the problem was that political pressures to spend would lead to high inflation. Here, economists proposed a solution in the form of "commitment devices," defined as: "institutional arrangements that make it a difficult and time-consuming process to change the policy rules in all but emergency situations" (Kydland and Prescott 1977: 487). In Rogoff's (1985) influential contribution, these arrangements were conceptualized as an independent central bank that would not be swayed by political considerations.

The idea that credible commitment to policies could be created by the delegation of decision-making to a politically insulated technocratic body subsequently spread across a range of policy areas in Europe and elsewhere. The resulting bodies were sometimes termed "non-majoritarian" institutions, that is, not reflecting the preferences of the majority of voters, and included arm's-length regulators and specialist agencies (Majone 2001; Thatcher and Stone Sweet 2002). In turn, arguments about depoliticization have been extended to the establishment of credible climate commitment (Helm et al. 2003; Brunner et al. 2012; Grosjean et al. 2016), where a now standard recommendation is the establishment of an independent advisory body for climate change policymaking, creating stability in climate policy trajectories (Averchenkova et al. 2018). At the core of this argument is the idea that delegation to an independent technocratic body removes distributional conflict from the political sphere (Meckling and Nahm 2018). This "depoliticizing" move is often conceptualized as the transfer of climate decision-making away from the realm of "formal politics" into more technocratic and less accountable spaces (see Paterson, Tobin, and VanDeveer, Chapter 1, this volume).

One critical response to this strategy has been to take the existence of such depoliticization at face value and draw attention to the kinds of democratic deficits that then arise (e.g. Wood 2016; for an application to energy and climate policy, see Kuzemko 2016). Another is to point to the tensions between arguments for stability through depoliticization on the one hand and those for the inevitability of political conflict in establishing and maintaining climate ambition on the other (Paterson et al. 2022). As this volume makes clear, depoliticization and stability are not clearly correlated properties of climate policy systems.

Here, however, we take a different approach by exploring the origins of these climate advisory institutions and questioning the degree to which they can be understood as apolitical in the first instance. We argue that framing climate advisory bodies as institutional efforts to depoliticize climate policymaking misrepresents the political dynamics that led to their creation. Existing work too often reads the instrumental effects of new climate institutions as the logic that motivated policymakers to propose and pass these reforms in the first place (e.g. Brunner et al. 2012). We instead argue that many climate advisory institutions are the path-dependent product of distributive and partisan conflicts, rather than efforts to bypass climate politics. These institutions did not emerge merely as a result of norms about public goods provision and efforts to reshape intertemporal policymaking incentives, to provide stability, or to solve the gap between current and future welfare needs. Instead, they often emerged to solve the short-term tactical needs of political actors. The subsequent effect of these institutions on the long time trajectory of climate policymaking was a by-product of political conflict, not an effort to sidestep that conflict. What seems in many cases like depoliticization – upon closer examination – proves anything but.

The origins of these climate institutions, in turn, shape their ability to create stability in national climate policymaking regimes (see Paterson, Tobin, and VanDeveer, Chapter 1, this volume). We argue that the political context surrounding their creation tends to limit the degree to which they can stabilize policy over time or depoliticize climate policy debates. We thus offer a distinct understanding of these climate institutions grounded in historical institutionalism rather than economic rational choice institutionalisms (Hall and Taylor 1996). This approach builds on the wider literature in politics that recognizes the possibility of multiple reasons for, and outcomes of, delegation (Flinders 2008).

We develop our argument with political readings of four cases of "depoliticizing" climate institutions. These comprise two countries (Australia, the UK) that created new bureaucratic bodies to delegate decisions on climate policy stringency and two countries (Norway, Denmark) that created new organizational venues to create long-term, cross-party policy commitments. We analyze each case, drawing from documentary evidence and eighteen interviews with senior political actors, bureaucrats, interest group advocates, and environmental leaders (see also Lockwood 2013, 2021a, 2021b; Mildenberger 2020).

In each case, we highlight the significant gap between the putative "depoliticizing" function of the country's climate institution and the fundamentally political origins of these institutions. These institutions were not deliberately established to circumvent politics but instead addressed the distributive core of climate politics, the short- or medium-term political needs of incumbent governments, or both. We then discuss the implications of how climate institutions were created and how they have subsequently functioned. We conclude by emphasizing the importance of correctly specifying the origins of climate institutions. When we conflate one possible institutional function with the dynamics of institution-building, we risk undermining efforts to generate the robust, new climate institutions necessary to address the climate crisis (e.g. Dubash et al. 2021).

13.1 The 2008 UK Climate Change Committee

The UK is often seen as the paradigmatic case of institutional innovation in climate policy, with the 2008 Climate Change Act and establishment of a Climate Change Committee (CCC). The act has been widely taken up and copied, in form if not always in content (Nash and Steurer 2019). Both the act and the CCC have been interpreted as mechanisms designed explicitly for strengthening the credibility of climate commitment (Brunner et al. 2012; Averchenkova et al. 2018). However, while this idea was discussed by some actors involved in its genesis, the act and especially the CCC can more accurately be seen as primarily the outcome of an episode of political party competition in the UK in the mid 2000s (Carter 2009; Lockwood 2013; Institute for Government 2012).

Before 2004, the salience of climate change was low in the UK and not subject to any party competition (Carter 2006; Carter and Childs 2018). On the contrary, the Treasury in the Labour government was wary of climate policy ambition after pushback in the late 1990s from energy-intensive industry and the road lobby, while the Conservatives were largely hostile. In early 2004, Labour Prime Minister Tony Blair launched a sustained campaign to raise the profile of the issue with a speech. However, this exposed the government, as its flagship climate change program from 2000 was widely acknowledged to be failing. Blair's intervention encouraged the civil society group Friends of the Earth to focus their campaign efforts on climate policy (Carter and Childs 2017), a move that was then picked up on by a wide range of actors, including churches and conservation organizations, as well as backbench MPs. Industry was still split (Mildenberger 2020), but leaders in key companies and the Confederation of British Industry were leaning in the direction of supporting greater ambition, reinforced by the publication of the "Stern Review," a high-profile government-commissioned assessment of the economic costs and benefits of mitigating climate change, in late 2005.

The situation then changed radically at end of 2005 when the Conservatives elected a new leader, David Cameron, who immediately sought to reposition the party in the middle ground, using climate change as the issue to do this. He visited the Arctic, changed the party symbol to a tree, and used the slogan "Vote Blue, Go Green" in the 2006 local elections (Carter 2009). The Conservatives and the Liberal Democrats signed a joint statement on a cross-party approach to climate change. Importantly, Cameron also publicly backed the civil society campaign for a new Climate Bill, which was getting wide support from backbenchers of all parties.

This strategy produced a large boost for the Conservatives in opinion polling, with Labour seriously challenged for the first time since 1997. The government came under huge pressure to respond – according to an aide, David Miliband, a high-profile Labour figure who as environment secretary was leading on climate policy, commented that "Labour could not get into the position of being the only major party not in favor of the proposed bill" (Institute for Government 2012: 115). Labour needed a solution and started looking at options in the second half of 2006 and formally announced its own bill in November 2006.

The origins of the CCC also lie in this process of political competition. The idea for an advisory and monitoring body did not come initially from the government, nor from the civil society campaign, which was instead very much focused on legally binding annual emissions reductions targets (Carter and Childs 2017). Instead, it was introduced by Cameron in an October 2005 speech (Tempest 2005), in the form of a proposal for a "Carbon Audit Office." It then appears in the cross-party climate consensus statement in January 2006. Institutional innovation was thus part of the agenda being set by political competitors to which the government had to respond. By mid 2006, aides to Miliband were discussing a proposed "Carbon Policy Committee" as part regulator and part monitoring and advisory body. However, the attraction lay as much in the blame avoidance function such a body could play as in the creation of credibility.

Thus, the origins of the act and CCC lie in an episode of sharp political competition, marking the emergence of Conservative challenge to Labour from which the latter party did not really recover. In the absence of this competition, it is unlikely that the act and the Committee would have been brought into existence.

13.2 The 2011 Australian Climate Change Authority

Few countries have seen more contentious debates over climate policy than Australia – or larger oscillations in policy outcomes (e.g. Cashore and Howlett 2007). During the 2007–2011 Labor government under Kevin Rudd, a high-profile effort to pass comprehensive climate reforms faltered and led to Rudd's replacement with new Labor leader Julia Gillard. After Julia Gillard won a surprise minority government in the 2011 federal election, she initiated a new climate policymaking effort in partnership with the Australian Green Party and several climate-minded independents that she depended on to govern. These negotiations took place within a newly established Multi-Party Climate Change Committee (MPCCC), which all parties were invited into, but which the right-wing opposition declined to join. The MPCCC served as a coordination venue for climate policymaking, bringing the minority Labor government into conversation with its political allies. The committee also included four nongovernment members, brought on for their individual expertise: a scientist, an economist, a social service provider, and an energy stakeholder. While these members did not have voting rights, they otherwise were included as equal members of the committee.

However, the MPCCC quickly encountered differences in the climate policy ambitions of an industry-aligned Labor government and more stringent voices from the Green Party. While both sides were able to negotiate over the structure of a national emissions trading scheme, they could not agree on the pollution reduction trajectory this scheme should have. The Greens refused to accept anything less than 25 percent reductions below 2000 levels, while Labor was reluctant to agree to anything more ambitious than 5 percent reductions below 2000 levels. Facing intractable conflict, the MPCCC instead agreed to establish a new independent agency, the Climate Change Authority (CCA) in July 2011, whose bureaucrats would be tasked with formally recommending the stringency of Australia's emissions trading scheme based on a set of jointly agreed-on criteria.

This move to delegate Australia's carbon pollution reduction trajectory generated a short-term consensus largely because the Greens and Labor had very different expectations about the CCA's likely recommendations. This stemmed from different interpretations of the CCA's mandate, which asked the CCA to calibrate

Australia's response based on the actions of other countries. Greens and Labor Party negotiators had different perspectives on what level of foreign climate policies would induce accelerated domestic pollution cuts. At the same time, the Labor government reserved the right to set the eventual target at its 5 percent level, overruling the CCA, if the Australian parliament could not vote to accept the CCA's report. In this sense, the CCA was a much weaker institution than its otherwise similar UK counterpart.

The CCA also proved weak because it could not survive Gillard's loss to the anti-climate Liberal Party in Australia's 2013 election. The MPCCC's proposed targets were ultimately released after Tony Abbott's election, surprising most Labor Party actors with its proposal of a 15 percent cut below 2005 levels – more ambitious than they had expected at the time of the CCA's creation. However, Abbott was already in the process of dismantling Australia's climate policy and the CCA's report was purely symbolic. Instead, climate policy retrenched in Australia despite the CCA.

Still, despite these failures, it is again instructive to consider the political origins of this move to delegate policy stringency decisions to an independent agency. Rather than reflecting a commitment to delegation or depoliticization, the CCA emerged as the only viable solution to intractable political conflict between Labor and the Greens during policy negotiations. The decision responded to short-term political needs and was established to meet these, rather than serve the type of credibility-creating functions that could be retrospectively read into the example.

13.3 The 2008 Norwegian Climate Settlement

A different type of seeming "depoliticization" of climate change occurred in Norway in 2008, with the country's Climate Settlement (*Klimaforliket*). These political agreements came in response to a government 2007 climate policy White Paper that brought all but one of the country's major political parties together to outline a shared and stable framework for medium-term national climate policymaking. Only the far-right Progress Party, which has downplayed the climate change threat, refused to participate in the settlement. This effort to establish a shared set of policy priorities that could outlast changes in governments would, on the surface, seem an archetypical example of politicians working together to solve the credible climate commitment problem. Rather than leaving climate decisions to oscillate as a function of political control, all major stakeholders came together to establish a shared decarbonization trajectory (amounting to between 13 million and 16 million metric tonnes) that could guide climate policy over the coming years. Moreover, such multiparty agreements have some precedent in Norwegian healthcare and pension policymaking (Mildenberger 2020).

However, the political motivation for establishing this Climate Settlement had little to do with political concern over long-term policy credibility. Instead, the settlement was a short-term political solution to an internal crisis within the incumbent governing coalition, led by Labour Prime Minister Jens Stoltenberg and his junior governing party, the Norwegian Socialist Left (SV) party. This de facto red–green coalition, elected in 2005, made early progress on such environmental files as biodiversity but struggled to navigate tensions over climate policy ambition. In particular, the industrial wing of the Labour Party refused to compromise with the strong environmental wing of SV, leading to significant internal friction that led SV politicians to contemplate leaving government. When Labour pushed through a climate policy White Paper in 2007 that was viewed as relatively unambitious, these tensions crystallized. SV found its government being criticized from the right by such parties as the Norwegian Liberals (Venstre) and Christian Democrats for the White Paper's willingness to allow Norway's carbon reductions to come from supporting decarbonization abroad rather than imposing costs on domestic industry (see Farstad, Hermansen, and Lahn, Chapter 11, this volume). SV found this position politically untenable.

Labour's solution to resolving this internal coalitional conflict was to expand the table to bring the opposition into the room. By including opposition parties in official climate policy development, they could defang the opposition's ability to destabilize Labour's partnership with SV. The ultimate settlement was only more ambitious than the White Paper and did not reflect any serious step change in Norwegian climate policy ambition. It remained significantly less ambitious than SV wanted and primarily innovated through its creation of a large-scale program to support avoided deforestation in the Global South. However, the settlement solved a political problem for the incumbent Labour–SV government, shielding the pro-climate SV from criticism by other political parties. In other words, the settlement was not motivated by a shared desire to create a credible reduction trajectory for the country. Instead, it was designed to protect a governing coalition from outside criticism.

13.4 The 2020 Danish Climate Act

The 2020 Danish Climate Act involves a combination of dynamics from the UK and the Norwegian case. A Social Democrat-led government had adopted an earlier Climate Change Act in 2014, which was superficially modeled on the UK's CCA and included a Climate Council (*Klimarådet*) to mirror the UK's CCC (Dyrhauge 2021; Lockwood 2021a). However, because attempts to gain opposition support were only partially successful, the Danish act lacked quantitative targets for emissions reduction or carbon budgets, gave a restrictive advisory remit to the Climate Council, and was seen as largely symbolic in nature (Nash and Steurer 2021).

By the late 2010s, the wider context had changed. Following the Paris Agreement, the salience of climate change in Denmark was rising sharply (Concito 2018; Dyrhauge 2021), and there was pressure from groups in civil society for a new Climate Change Act with quantitative targets and an interim planning system for reaching these (Nash and Steurer 2021). The center-right government in power by this time did not, however, manage to negotiate another climate agreement to form the political basis for a new act before the 2019 election.

The issue-attention cycle on climate change made the 2019 election in part a "climate election," and there was political competition over the issue, especially among the left parties (the Socialist People's Party [SPP], Social Liberals, and Red–Green Alliance) that led these to adopt the civil society proposals in full. However, the 2019 election was also about immigration (Green-Pedersen 2020), where the main center-left party, the Social Democrats, had moved to the right, to compete for votes from the populist Danish People's Party (DPP) (Kosiara-Pedersen 2020). To convince voters that they would not abandon their hardline stance on immigration in postelection negotiations with the left parties, the Social Democrats announced beforehand that they would form a minority government if elected with the largest share. In the event, this was indeed the outcome, with the DPP vote collapsing.

However, this electoral shift put the Social Democrats in a difficult position; to govern they still needed the support of parties on the left but they had stood on a platform of restricting immigration and promoting integration that those parties, especially the SPP and the Red–Green Alliance, opposed. Negotiations took twenty days, the longest period since 1988, complicated by the need to reach a trade-off between changes to the Social Democrats' immigration policy and the left parties' demands on climate targets (Selsoe Sorensen 2019). The resulting platform meant the Social Democrats had to accept a stronger proposed agreement, making Denmark "one of the most ambitious countries in the world on climate change,"[1] but gave it a basis to govern. A consequent Climate Agreement was reached with eight out of ten parties in December 2019, followed by a new Climate Change Act in 2020 with the new targets, a planning review system, a stronger role for the Climate Council, and greater parliamentary oversight (Lockwood 2021a; Nash and Steurer 2021).

As in the UK in the 2000s, a wave of heightened salience of climate change created party-political competition but created a problem for the Social Democrats who were as much, if not more, focused on competition over immigration with the DPP. Once elected as the largest single party, but committed to forming a minority government, they faced a short-term political problem of how to gain parliamentary support without giving way on their immigration policy position. The

[1] Maria Reumert Gjerding, Danish Society for Nature Conservation, quoted in Henley (2019).

resulting Climate Agreement and Climate Change Act proposals, stronger than those that the center parties would have wanted, can thus be seen as the outcome of a solution to that problem.

13.5 The Illusion of Apolitical Climate Policy

Across the world, a handful of climate institutions have been held up as possible models for a new apolitical approach to addressing the climate crisis. Here, we have suggested that – upon closer empirical reexamination and comparison – four of the most prominent global examples of "depoliticizing" climate institutions proved anything but. In each case, these institutions did not respond to a desire to circumvent politics or enhance credible commitments. Instead, these institutions helped bridge otherwise intractable conflicts or protected incumbent governments from political controversy. Often, they resulted from a stalemate over policy stringency between different coalition or government actors. These cases highlight at least three critical considerations when it comes to the analysis of climate institutions: the importance of giving attention to the political conditions that led to the creation of these new institutions; the limited capacity of these institutions to transcend these initial political conditions; and the broader diversity of institutional solutions that will likely be necessary to overcome the challenging politics of climate change.

First, climate institutions – including institutions that provide the appearance of depoliticizing climate policy – have to be understood in the context of how they were created. Notwithstanding the eventual functions that these "depoliticizing" institutions have played in each of these four countries, a careful unpacking of their origins reveals a gap between the political motivations that established them and these potential functions. For example, in both Norway and Australia, the delegation of climate policymaking decisions to either an independent agency or cross-party agreement did not reflect desire to address long-term problems of credible commitment and policy stability. Instead, these institutions provided short-term political benefits to incumbent governments, helping them to navigate contentious political conflicts. Similarly, in the UK both the Climate Change Act and the accompanying institutional innovation of the Climate Change Committee and in Denmark the 2020 Climate Agreement and Climate Act were driven by episodes of intense party-political competition.

The issue, of course, is that a theory of institutional origins that privileges potential long-term functions and a theory that focuses on short-term distributive-political drivers can yield observationally equivalent institutional designs. In other words, the outcomes of these distinct institution-building stories can be similar, even if they have radically different operational logics. This creates a premium on empirical study of the motivations behind climate institution-building, to better predict and support the conditions under which subsequent climate institutions can emerge (Dubash et al. 2021).

Second, the origins of these institutions have significant implications for institutional efficacy and policy stability over time. Because the creation of the four institutions considered here was driven primarily by the need to solve immediate political problems rather than long-term policy problems, their design was typically based on a set of compromises and they are often relatively weak, that is, they usually lack the decision-making authority seen in central banks and other non-majoritarian bodies (i.e. regulators). Such powers were explicitly ruled out in the case of the UK's much-celebrated CCC, for example (Lockwood 2013, 2021b). As a result, the repoliticization of climate policy, in the sense of partisan or distributive conflict, or both, is inevitable (and, indeed, is a misnomer of sorts since the institutions did not necessarily accomplish meaningful depoliticization in our account).

Moreover, as Posen (2010) points out, even the effectiveness of independent central banks to whom decisions on interest rates are delegated depends ultimately on political support from others:

> The only way that central banks can credibly commit to price stability over the long-term is to maintain a political constituency in civil society supportive of such a policy regime. That support from civil society, not any legal statute, is what protects central banks when they make a hard decision that angers politicians. Absent that support, laws regarding central banks can be changed or threatened to be changed until monetary policy is changed. Central bank independence is endogenous to that support, and it will be curtailed when such support is lost. (Posen 2010: 5)

The same is arguably even more true for the climate domain. The relative weakness of credible climate commitment institutions matters because, while they may attract less attention from opponents and are thus more likely to survive, it also means that they are less likely to produce positive feedback effects. They are likely to leave the actors, conflict between whom originally produced them, untransformed. Institutions are unlikely to become a route to circumvent politics. Instead, we should expect that institutions are the route through which politics will continue to structure climate policymaking over the coming decades.

A reasonable objection here is to consider whether the logic of credible commitment and the logic of blame avoidance (e.g. responsibility for imposing costs on voters) are functionally different (Flinders 2008). One might argue that the mechanism of a delegation either to a technocratic body or to a political agreement have similar tactical justifications. For instance, the Canadian province of New Brunswick established a multiparty select committee to plan its carbon price in 2016 (Mildenberger 2020). Here, the intent was to obscure responsibility for potential costs resulting from the policy. Still, in at least the Australian and Norwegian cases, we see political maneuvering that is clearly more oriented toward handling short-term conflicts. By contrast, the UK case has a slightly different orientation. In the lead-up to the Climate Change Act, there was a supportive wind of public opinion for stronger action, external policy entrepreneurs pushing for a long-term framework, and a resurgent opposition making climate change central to its platform. This situation created a problem of party-political competition for action for the government, in the context of the widely held perception that its climate change program of the early 2000s had failed. In Denmark, all political parties were under pressure from civil society and public opinion to raise ambition, but the outcome was again determined by short-term political trade-offs with the immigration issue agenda in the 2019 election.

If the institutions we discuss here were created purely for the purposes of meeting climate commitment functions, we would expect to see consensual or technocratic policymaking, or at least that conflicts which arose after they came into being would be managed and resolved by those institutions. However, in some cases we find the opposite, that is, that when short-term political conflicts over climate policy resurface, they are played out outside the architecture of credible commitment. This is especially the case for delegation approaches (Lockwood 2021a). For example, when in 2013 there was conflict between the Conservatives and the Liberal Democrats in the UK about the setting of the fourth carbon budget (with business and environmentalist campaign groups pitching in), this was resolved not by the Climate Change Committee but rather by an informal negotiation process between party leaders (Lockwood 2013, 2021b). The same is the case in Australia. While delegation to the CCA initially allowed for management of the distributive conflict between otherwise aligned Labor and Green Party political actors, it obviously could not protect the country's carbon pricing system from retrenchment when the anti-climate Liberal government came into power in 2015.

Third, and finally, these political dynamics remind analysts, advocates, and policymakers that a richer institutional landscape will be necessary to address climate policy needs. There are no institutional silver bullets for managing the complex economic and social conflict implied by decarbonization. At one level, while all four of our country examples involved institutional innovation, they differed in nature. Australia and the UK adopted the standard recommendation from institutional economics for addressing the credible commitment problem by setting up independent non-majoritarian bodies. By contrast, in Norway and Denmark, political parties brokered consensus-based agreements or settlements. This contrast arguably reflects different wider institutional contexts, notably proportional representation electoral rules and coalition politics in the two Scandinavian cases (Lockwood 2021a), in contrast to the majoritarian systems in Australia and the UK.

Yet even this variation exists within a narrow range of institutions that delegate or change the venue for decision-making. Instead, institutions that happen to solve credible commitment problems are not the only ones – nor necessarily the best ones – to solve distributive-institutional conflict. We do not see anything universal about "credible commitment" institutions as the recipe for solving climate policy conflict – conflict that instead varies across countries and demands tailored institutional responses. Consider the case of the United States, which instead pursued a deliberate policy of rejecting credible commitment policy features during the Obama administration. In particular, Environmental Protection Agency (EPA) bureaucrats chose a policy of what might be characterized as maximum threat with uncertainty (Mildenberger 2020). They felt that the best way to achieve real transitions in investment structure was to create uncertainty about policy ambition over medium time horizons. Here, the Clean Power Plan is instructive: Its EPA champions viewed it as having an uncertain future before the US Supreme Court. However, rather than trying to pass a policy that was viewed as legally durable, advocates felt it was more important to threaten a more ambitious – even if uncertain – framework to reshape business expectations about long-term investment payoffs. At least one architect of the policy expressed agnosticism about whether the Clean Power Plan would come into effect (Mildenberger 2020). Instead, simply the proposal – even under review by the courts – could impact business perceptions of future policymaking environments and serve its purpose in the absence of coming into effect.

In sum, prominent "depoliticizing" climate institutions have been misunderstood. Their apolitical nature is illusory and misinterprets their origins, their potential impacts, and their contributions to supporting climate policy stability through time. It is tempting to wish for institutional innovations that can circumvent the distributive politics of climate change. To date, however, there are no institutional ways to bypass climate policy conflict. Instead, climate advocates and policymakers have no choice but to recognize and engage with this conflict and work through it.

References

Averchenkova, A., Fankhauser, F., and Finnegan, J. (2018). The role of independent bodies in climate governance: The UK's Committee on Climate Change. Grantham Research Institute on Climate Change and the Environment/Centre for Climate Change Economics and Policy. www.lse.ac.uk/GranthamInstitute/wp-content/uploads/2018/10/The-role-of-independent-bodies-in-climategovernance-the-UKs-Committee-on-Climate-Change_Averchenkova-et-al.pdf

Bosetti, V., and Victor, D. G. (2011). Politics and economics of second-best regulation of greenhouse gases: The importance of regulatory credibility. *The Energy Journal* 32(1), 1–24.

Brunner, S., Flachsland, C., and Marschinski, R. (2012). Credible commitment in carbon policy. *Climate Policy* 12(2), 255–271.

Carter, N. (2006). Party politicization of the environment in Britain. *Party Politics* 12, 6.

Carter, N. (2009). Vote blue, go green? Cameron's conservatives and the environment. *Political Quarterly* 80(2), 233–242.

Carter, N., and Childs, M. (2018). Friends of the Earth as a policy entrepreneur: "The Big Ask" campaign for a UK climate change act. *Environmental Politics* 27(6), 994–1013.

Cashore, B., and Howlett, M. (2007). Punctuating which equilibrium? Understanding thermostatic policy dynamics in Pacific Northwest forestry. *American Journal of Political Science* 51(3), 532–551.

Concito. (2018). Klimabarometeret 2018. https://concito.dk/sites/concito.dk/files/dokumenter/artikler/klimabarometeret_2018_endelig_rev1.pdf

Dubash, N. K., Pillai, A. V., Flachsland, C. et al. (2021). National climate institutions complement targets and policies. *Science* 374(6568), 690–693.

Dudley, H., Jordan, A., and Lorenzoni, I. (2021). Independent expert advisory bodies facilitate ambitious climate policy responses. ScienceBrief Review.

Dyrhauge, H. (2021). Political myths in climate leadership: The case of Danish climate and energy pioneership. *Scandinavian Political Studies* 44(1), 13–33.

Flinders, M. (2008). *Walking without Order: Delegated Governance and the British State*. Oxford: Oxford University Press.

Green-Pedersen, C. (2020). Issue competition in the 2019 Danish general election. *Scandinavian Political Studies* 43(2), 90–95.

Grosjean, G., Acworth, W., Flachsland, C., and Marschinski, R. (2016). After monetary policy, climate policy: Is delegation the key to EU ETS reform? *Climate Policy* 16(1), 1–25.

Hall, P. A., and Taylor, R. C. R. (1996). Political science and the three new institutionalisms. *Political Studies* 44(5), 936–957.

Helm, D., Hepburn, C., and Mash, R. (2003). Credible carbon policy. *Oxford Review of Economic Policy* 19(3), 438–450.

Henley, J. (2019). Denmark's youngest prime minister to lead new leftist government. *The Guardian*, July 26. www.theguardian.com/world/2019/jun/26/denmarks-youngest-prime-minister-leads-new-leftist-government

Institute for Government. (2012). The Climate Change Act (2012). www.institutefor government.org.uk/sites/default/files/climate_change_act.pdf

Jacobs, A. (2011). *Governing for the Long Term: Democracy and the Politics of Investment*. Cambridge: Cambridge University Press.

Kosiara-Pedersen, K. (2020). Stronger core, weaker fringes: The Danish general election 2019. *West European Politics* 43(4), 1011–1022.

Knaggård, Å., and Pihl, H. (2015). The green state and the design of self-binding. In K. Bäckstrand and A. Kronsell, eds., *Rethinking the Green State: Environmental Governance towards Climate and Sustainability Transitions*. London: Routledge, pp. 209–224.

Kuzemko, C. (2016). Energy depoliticization in the UK: Destroying political capacity. *BJPIR* 18(1), 107–124.

Kydland, F. E., and Prescott, E. C. (1977). Rules rather than discretion: The inconsistency of optimal plans. *Journal of Political Economy* 85(3), 473–492.

Lazarus, R. J. (2009). Super wicked problems and climate change: Restraining the present to liberate the future. *Cornell Law Review* 94, 1153–1234.

Lockwood, M. (2013). The political sustainability of climate policy: The case of the UK Climate Change Act. *Global Environmental Change* 23(5), 1339–1348.

Lockwood, M. (2021a). Routes to credible climate commitment: The UK and Denmark compared. *Climate Policy* 21(9), 1234–1247.

Lockwood, M. (2021b). A hard act to follow? The evolution and performance of UK climate governance. *Environmental Politics* 30(Supp 1), 26–48.

Majone, G. (2001). Non-majoritarian institutions and the limits of democratic governance: A political transaction-cost approach. *Journal of Institutional and Theoretical Economics* 157(1), 57–78.

Meckling, J., and Nahm, J. (2018). The power of process: State capacity and climate policy. *Governance* 31(4), 741–757.

Mildenberger, M. (2020). *Carbon Captured.* Cambridge, MA: MIT Press.

Nash, S., and Steurer, R. (2019). Taking stock of climate change acts in Europe: Living policy processes or symbolic gestures? *Climate Policy* 19(8), 1052–1065.

Nash, S., and Steurer, R. (2021). From symbolism to substance: What the renewal of the Danish climate change act tells us about the driving forces behind policy change. *Environmental Politics* 31(3), 453–447.

Paterson, M., Tobin, P., and VanDeveer, S. D. (2022). Climate governance antagonisms: Policy stability and repoliticization. *Global Environmental Politics* 22(2), 1–11.

Posen, A. (2010). When central banks buy bonds: Independence and the power to say no. Speech presented at Barclays Capital 14th Annual Global Inflation-Linked Conference, New York, June 14. www.bankofengland.co.uk/-/media/boe/files/speech/2010/when-central-banks-buy-bonds-speech-by-adam-posen

Rogoff, K. (1985). The optimal degree of commitment to an intermediate monetary target. *Quarterly Journal of Economics* 100(4), 1169–1189.

Selsoe Sorensen, M. (2019). Denmark's new prime minister vows to tackle climate change. *New York Times*, June 26. www.nytimes.com/2019/06/26/world/europe/denmark-prime-minister-mette-frederiksen.html

Tempest, M. (2005). Cameron calls for climate change consensus. *The Guardian*, October 31. www.theguardian.com/environment/2005/oct/31/toryleader.politics

Thatcher, M., and Stone Sweet, A. (2002). Theory and practice of delegation to non-majoritarian institutions. *West European Politics* 25(1), 1–22. https://doi.org/10.1080/713601583

Wood, M. (2016). Politicization, depoliticization and anti-politics: Towards a multilevel research agenda. *Political Studies Review* 14(4), 521–533.

Part IV

Global Politics

14

Politicization Conflicts in Global Climate Governance

JEN IRIS ALLAN

Climate change was introduced to the global agenda in the late 1980s. Over the next three and a half decades, countries negotiated and renegotiated the rules. To an extent, the need to revisit and reset the rules in the global climate regime stems from conflicts between actors seeking to repoliticize and depoliticize the resulting treaties. Efforts to question and rewrite rules require as much diplomatic and material investment as maintaining stability through depoliticizing potentially contentious topics. While the literature tends to focus on one dynamic or the other, there is a need to understand how these two opposing coalitions interact (Paterson, Tobin, and VanDeveer 2022). As this chapter shows, these tensions play out in unique ways at the global level, where stability can be fragile and difficult to maintain. Politicization, on the other hand, can come from many sources and employ a range of behavioral and rhetorical tools.

This chapter focuses on one such conflict, when the United States sought to repoliticize the Kyoto Protocol and the European Union (EU) sought to protect it by depoliticizing discussions. Each drew upon a range of strategies, rhetorical and behavioral, creating a rift between the countries. Eventually, the EU and the United States accommodated one another, changing the course of the climate regime. New concepts became mainstream, notably a "bottom-up" treaty allowing all countries to pledge their own targets. The Kyoto Protocol was set aside to make room for a new agreement and US involvement. These changes masked stability as policy lock-in. The United Nations Framework Convention on Climate Change (UNFCCC) was solidified as the central forum for climate governance. Market mechanisms remained central. Even the mitigation rules showed institutional stability despite the rhetorical changes.

A central subject of repoliticization and depoliticization efforts since the 1990s can be broadly termed "burden sharing," or which countries should do the most to reduce their emissions. Mitigation rules vary somewhat among the three treaties in the climate regime: the UNFCCC (adopted in 1992), the Kyoto Protocol (adopted in 1997), and the Paris Agreement (adopted in 2015). The UNFCCC identifies industrialized countries and economies in transition as "Annex I" countries and specifies several principles related to "developed country leadership" and "common but differentiated responsibilities and respective capabilities" (CBDR-RC). The Kyoto Protocol specifies that only Annex I countries have legally binding emissions targets, while non-Annex I (developing) countries can undertake voluntary measures. The 2015 Paris Agreement requires all countries to submit nationally determined contributions (NDCs). Developed countries are to have quantitative targets, while developing countries have more flexibility; they can submit plans or policies, for example. The question of burden sharing has expanded to include all countries in the mitigation effort, although with flexibilities allowed by the nationally determined (or "bottom-up," but for a critique of this label, see Depledge 2022) nature of the Paris Agreement.

There has been considerable change to expand mitigation rules to all countries and to soften the hard distinctions between Annex I and non-Annex I into newer categories or developed and developing countries. Yet there are several similarities between the Paris Agreement and the 1997 Kyoto Protocol. Both agreements share similar categories of climate policy (except for "loss and damage" that appears only on post-Kyoto agendas) (Allan and Bhandary 2022). Both give a central role to market mechanisms that allow emissions trading. The substantive mitigation requirements of both agreements involve nationally determined target setting (Depledge 2022). Countries pledge their NDCs under the Paris Agreement, while the Kyoto Protocol famously inscribed the emissions reduction targets for developed countries into the treaty. However, the process to arrive at the Kyoto targets was unilaterally determined – in the final hours, developed countries wrote down their targets and handed the papers to the Secretariat without further multilateral discussion (Depledge 2022; UNFCCC 2000). National determination and market mechanisms, to name a few, seem baked into the climate regime.

This chapter outlines politicization strategies used in international climate negotiations, and how they played out in the case of the Kyoto Protocol. There are key lessons to be learned for contemporary climate politics. There are signs that the United States, this time with the EU, is repoliticizing the Paris Agreement. This time, China, India, and other larger developing countries are engaging in similar depoliticization strategies used by the EU in the 2000s. This may again prolong the implementation of climate policies while countries debate how to move forward.

14.1 Politicization Strategies in International Climate Negotiations

As Chapter 1 outlined, there is a push–pull relationship between depoliticization and repoliticization. Depoliticization involves actors seeking to remove an issue from the spotlight, often by framing it as a technical exercise. Repoliticization, by contrast, is characterized by efforts to add an issue to a political or public agenda, highlight power imbalances, and urge continued debate. These strategies can occur simultaneously, over the same rules and issues, leading to conflict or compromise.

The unique characteristics of global governance influence how politicization conflicts unfold. Unlike in domestic politics, the stability of global politics is more or less secure. Global rules are upheld because countries agree to be bound to them. Stability rests on legitimacy and countries can withdraw their consent. Beyond states, there is a wide range of actors that can exert influence over global outcomes, formally and informally. These actors meet yearly to review policies' implementation, resume negotiations, or raise new issues. There is no electoral cycle, when policies could be more stable if implemented by a governing party or coalition. At the global level, many can regularly contest the rules.

Many actors could repoliticize or depoliticize an issue or rule in global environmental politics. It is not necessarily the case that the weak seek to expand the scope of conflict while the powerful seek to limit it, as Schattschneider (1960) suggests. States are legally equal. In the UNFCCC, decisions require consensus. Some states, notably the United States, clearly pull considerable weight, shaping global rules to suit domestic legal needs (Kemp 2016). But also, small island states and other climate-vulnerable countries exert moral authority leading to influence that outstrips their economic power (de Águeda Corneloup and Mol 2014). In a situation where many actors wield various types of authority, repoliticization could come from a range of actors working together formally or tacitly.

Repoliticization can occur at any time because there are routine opportunities to politicize global rules and to reinterpret policy design. Global negotiation fora are key sites of agreement-making where actors advance, resist, or alter how we understand or govern key climate issues (Hughes et al. 2021). Agenda-setting politics are often thought to occur when negotiations are in their infancy. States can propose new agenda items. Such proposals are often hotly contested as parties debate the mandate, scope, and name of the new agenda item; many ultimately cannot gain the consensus of all states and fail to make it to the formal agenda (Allan and Bhandary 2022). For example, calls for a Loss and Damage Fund in 2021 sparked a year of debate and agenda-setting efforts to get the issue onto the formal UNFCCC agenda in 2022. Those negotiations established a transitional committee, with membership limited to fourteen countries. The group had a tight mandate to recommend institutional arrangements, based on an analysis of existing funding sources, gaps, and potential sources (UNFCCC 2022). Both repoliticization and depoliticization were at play. Developing countries set a new agenda, resisted by developed countries who sought to depoliticize the issue by limiting the scope of discussion to technical discussions outside of major meetings.

Even after the agenda-setting stage, the annual (or biennial) meetings of parties to a treaty provide regular opportunities for repoliticization using a range of strategies. A key strategy could be to delegitimate the rules, even to the extent of upturning international agreement entirely. Rhetorically, actors could question the efficacy or desirability of existing rules and provide alternatives to the status quo. Repoliticization efforts can seek to reframe existing issues as unjust or inefficient.

In addition to these discursive tactics, actors can use actions to raise questions about existing issues or rules. States can leave a treaty or refuse to join to undermine how others perceive it and its overall efficacy (Bäckstrand and Söderbaum 2018). Also unique to the global level, states can create alternative fora to discuss and advance alternatives. "Minilateral" solutions to climate change could be more efficient, by reducing the numbers of those involved. But by definition, they lack inclusion (Eckersley 2012; Gampfer 2016). At the very least, alternative fora help test or diffuse ideas among other states, without the constraints of rules of procedure and history of other fora. New fora could be an attractive strategy because there are fewer opportunities for repoliticization within the UNFCCC negotiations.

However, upholding the status quo requires responding to repoliticization while maintaining support coalitions. Maintaining legitimacy, after all, requires the consent of all or most states. There are several options. Technical negotiations abound in global environmental politics. Such discussions can sort through technical issues – for example, the finer points of global monitoring networks. But often, deeply political issues are repackaged as technical. Discussions regarding market mechanisms require decisions over which countries are eligible, how to ensure emissions reductions, and who receives the benefits (financial and carbon credit), among many other political issues masked through debates of double counting, CO_2 equivalence, and reference levels. As the Loss and Damage Fund example shows, negotiations can lead even the most political discussions toward seemingly technical, limited debate.

Relatedly, limiting the scope of discussions is an additional tactic to depoliticize global negotiations. Countries can limit the time for a work program or set a deadline for discussions. Work programs or dialogues are common ways to do this. Usually two or three years long, such road maps often set out a series of workshops or roundtable discussions. The mandate for such work programs often seeks to support, rather than overturn, the status quo. For example, the Glasgow–Sharm El Sheikh work program on the global goal for adaptation seeks to, among other objectives, "enable the full and sustained implementation of the Paris Agreement, towards achieving the global goal on adaptation, with a view to enhancing adaptation action and support" (UNFCCC 2021).[1] The work program on long-term

[1] The other seven objectives also support a limited view of the work on the global goal on adaptation, including to "enhance understanding," "contribute to reviewing progress," "enhance national planning and implementation," and "enable parties to better communicate their adaptation needs."

finance has featured biennial reports on countries' efforts to scale up reports, annual workshops, and biennial high-level dialogues (UNFCCC n.d.). As yet, no tangible outcomes are apparent, and after several years of debate, countries agreed to conclude the work program in 2027.

Beyond negotiation tactics to take issues out of the spotlight and limit the scope of potential reforms, legitimacy can be a resource aiding depoliticization. Countries, non-state actors, and others can point to hard-fought, globally agreed rules as something to protect and uphold. Citing these rules, and the legitimacy they may hold by those who agreed to them, can be an important defense against repoliticization.

Repoliticization and depoliticization strategies work in opposite directions. Where repoliticization strategies seek to enlarge the scope of conflict and prompt change, depoliticization seeks policy stability. When these strategies collide, hard bargaining and collective meaning-(re)making ensue. The repoliticization camp employs tools to build support for new systems, while the depoliticization camp remains invested in the existing policies. Perhaps surprisingly, there is room for compromise. Amending an existing policy and making it seem new could satisfy both sides. The rhetoric of repoliticization is adopted, which creates new opportunities for change. The underlying policies remain largely stable, rewarding the investments of those supporting the status quo.

In global climate politics, politicization debates are most evident around questions of burden sharing, or "who does how much." Usually, burden sharing debates occur in the context of mitigation. After transparency, mitigation is the second-most discussed topic on UNFCCC agendas, although half of those discussions are buried in technical talks related to markets and forests (Allan and Bhandary 2022). Countries long worried about comparative economic advantage have sought to ensure all their economic rivals have the same obligations. Some have refused participation at times (e.g. the United States). In other cases, developed countries raised concerns about emerging economies, such as China and India. Arguments and various notions of "fairness" abound.

14.2 Politicization Conflicts over the Kyoto Protocol

In the early years, the focus of the regime was industrialized countries' emissions. The UNFCCC, adopted in 1992, set out several rules and principles that would be the focus of politicization efforts for decades. Each of these rules and principles centers around which countries should be primarily responsible for reducing emissions or "burden sharing" in the UNFCCC's lingo.

The commitments set out in the Convention set out a differentiated approach. In the early 1990s, this was less controversial. There were clearer divisions between developed and developing countries before the rapid economic growth of some non-Annex I countries that had a largely taken-for-granted status. Dividing countries into two groups was institutionalized in the Convention. Broadly, industrialized countries were listed in Annex I of the Convention, and all other countries became "non-Annex I." Responsibilities vary between these two groups.

The Kyoto Protocol upheld the bifurcated approach to burden sharing. It inscribes legally binding targets and timelines into the treaty itself, but only for Annex I countries. This was less of an imposition on Annex I countries than it may seem. As Depledge (2022) explains, countries chose their targets. There were negotiations, but at the end of the long discussions, negotiators from Annex I countries wrote their targets on paper and handed them to the Secretariat for compilation and adoption without further discussion (Depledge 2022). The Protocol encourages non-Annex I countries to take on mitigation actions voluntarily.

The bifurcated approach was, and in some ways remains, at the center of repoliticization efforts. The United States questioned the feasibility and efficacy of the Kyoto Protocol even before it entered into force, pointing to its domestic requirements for a treaty to "meaningfully" engage emerging economies, and continued difficult international negotiations (Lisowski 2002). The EU sought to uphold the Protocol, given its many investments in the Protocol's institutions.

14.2.1 Depoliticization Strategies

For many, the Kyoto Protocol was a significant global achievement to be protected and implemented. The EU, international NGOs, and other countries invested materially, diplomatically, and discursively into ensuring the Protocol entered into force. They also invested in its institutions, building global and regional carbon markets, improving science around the carbon sink potential of forests, and reducing emissions to meet the Protocol's targets.

The EU was a staunch advocate for the Protocol. It invested heavily to secure leadership in global climate politics (Christiansen and Wettestad 2003; Gupta and Ringius 2001; Paterson 2009). It pursued a strategy to implement the Kyoto Protocol without the United States. It viewed its strategy as "walking the walk" to bring countries along. In particular, the EU invested heavily into making carbon markets work. The EU Emissions Trading Scheme was developed and linked to the Protocol's Clean Development Mechanism (CDM). This strategy involved reducing its own emissions, which, some argue, decreased the price of renewables, helping to make the case to developing countries that climate action was in their interest (Betts 2021: 114). With developing countries, the EU became a major source of climate finance. It used the CDM to build trade and investment relationships with China and India (Paterson 2009). The strategy also aimed to boost material investment and ideational support for the Kyoto Protocol.

Diplomatic levers were a common strategy to ensure the continuation of the Protocol. The EU worked to secure Russia's ratification by leveraging World Trade Organization membership (Paterson 2009). EU leaders hoped President George W. Bush could also be brought on board. Shortly after his election in 2001, EU member states invited him to a meeting in Sweden that ended in a stalemate (Black and White 2001). Ten days later, the European Council "reaffirm[ed] its commitment to delivering on Kyoto targets and the realization by 2005 of demonstrable progress in achieving these commitments" (EU 2001: 8). Tense EU-US relations on climate change characterized the 2000s, forging the "transatlantic climate divide" (Schreurs 2004; Szarka 2012). Regardless of the American position, the EU was firmly committed to its leadership role.

Within the UNFCCC negotiations, the EU pushed for the conclusion of technical negotiations. These negotiations sought to work out the operational rules around the Protocol's provisions, particularly on forests and markets. The EU had a strong interest in ensuring these aligned with its domestic policies. Somewhat awkwardly, the United States continued to provide input to these operational negotiations around the Protocol, despite not ratifying the Protocol. This did not seem to faze the EU, which continued to lead these talks (Paterson 2009). When the technical rules were adopted in 2001, this hope was still alive, as Conference of the Parties (COP) President Jan Pronk, from the Netherlands, remarked that the outcome "now provides the US with a suitable legal structure to join the process of combating global climate change" (IISD 2001).

The EU wanted legally binding emissions reduction targets for the United States, China, India, and other major economies. Its first preference was to accomplish this by amending the Kyoto Protocol's architecture; the second choice was a new treaty that looked very much like the Protocol (Kulovesi and Recio 2023). The Protocol's entry into force provided an opportunity to support and augment the existing rules and test other countries' willingness to sign up for mitigation targets. At the 2005 climate meeting in Montreal (COP11), the EU pushed for further technical work under the Protocol's Articles 3.9 (Annex I countries' targets in subsequent commitment periods, i.e. after 2012) and 9 (periodic reviews of the Protocol). The broader Article 9 review, the EU contended, should include all provisions and decisions taken under the Protocol to date (UNFCCC 2006). While Japan and Australia specifically mentioned the need to bring all major emitters into the Protocol through the review, the EU stressed that "enhanced cooperation" was necessary in light of the science. Calling the Protocol "innovative," the EU also preferred a focus on markets, forests and land use, aviation and shipping emissions, and sources and sectors (UNFCCC 2006).

There was also a strategy for the second option, a new treaty that looked much like the Kyoto Protocol. In Montreal, the EU pushed for the Convention Dialogue. It was a two-year series of roundtable discussions on "long-term cooperative action to address climate change by enhancing the implementation of the Convention" focused on advancing development goals sustainably, adaptation, technology, and markets (UNFCCC 2005). After the United States left negotiations on this dialogue because it worried that new commitments would be on the table, the EU could better push through its agenda (IISD 2005). The EU focused on existing rules, and largely sought to frame the discussions in technical terms. The goal was to operationalize and implement the existing rules, not create new ones.

The support for the Protocol and the Convention inside the negotiations was echoed by the EU's diplomatic engagement through the G8. The 2005 G8 meeting in Gleneagles, Scotland, was one of the first with dedicated, high-profile discussions related to climate change. UK Prime Minister Blair set the tone before the summit, stating that the aim of the talks would be how to move forward in 2012 when the Kyoto Protocol's first commitment period expired (Geoffrey and Procter 2005). As France's President Jacques Chirac put it:

> The entry into force of the Kyoto Protocol is a historic step. But we also know that we will have to go much further. The first commitment period for reducing CO_2 emissions from developed countries ends in 2012. We must prepare for the follow-up, within the UNFCCC, of which we are all members … We will have to begin at the Montreal conference, happening at the end of this year, negotiating a new mandate, the first step towards a future international framework for combating climate change that fully integrates development objectives. (G8 2005)

The wording was significant. The "follow-up" or "new international framework" could be a second commitment period or a new treaty. Several leaders carefully separated the Protocol from the UNFCCC. It strongly supported the current institutions the EU had already spent considerable resources to uphold. The strategy also left the door open to American involvement. That hope increased with President Obama's election and a Democratic majority in both houses of Congress. EU member states and negotiators saw the opportunity for a Kyoto Protocol-style agreement that would apply to all, have strong accountability provisions, and generate ambitious deep emission reductions (Betts 2021: 115).

Throughout the 2000s, the EU was a global leader, strongly supporting the existing global responses to climate change. Within the UNFCCC negotiations, the focus was on ensuring the Protocol entered into force and continued technical negotiations on markets and forests. When the opportunity arose to potentially expand or amend the existing rules through the reviews of the Protocol and new discussions on future global arrangements, the EU framed these firmly within the global architecture of the Protocol and the Convention.

14.2.2 Repoliticizing the Protocol

The US strategy to repoliticize the Kyoto Protocol took two forms. Inside the negotiations, the United States limited its engagement. Outside the negotiations, there was considerably more action. The United States built support for an alternative form of cooperation through new fora, drawing on rhetorical tools originating from US-based think tanks. It also openly questioned the Protocol's logic and efficacy.

Under the Clinton administration, the United States secured several wins in the Kyoto Protocol negotiations, including flexible emissions accounting and market mechanisms. Yet the United States knew it was negotiating a treaty that would be nearly impossible to ratify at home. In 1997, as the Protocol was under negotiation, the Byrd–Hagel Amendment passed in the United States with rare bipartisan support. It stressed that the United States would only ratify a climate treaty with wider participation by emerging economies (Harrison 2010). Without such "meaningful participation," the world's largest emitter failed to ratify the treaty. The United States underlined domestic worries about the effects of the Kyoto Protocol on the economy and votes (Harrison 2010).

Walking away from the Kyoto Protocol was a behavioral strategy to delegitimize it. The George W. Bush administration seemingly believed that this strategy would be enough to undermine the treaty, possibly prompting a new direction. After the meeting with the EU in 2001, where the EU hoped to bring the United States on board, US Secretary of State Condoleezza Rice countered, "Kyoto is dead" (Paterson 2009).

The administration pointed to several reasons why the United States would not join the global effort. Bush declared the treaty "fundamentally flawed" and a threat to American jobs. It was largely rhetorical. Actual US climate policy remained similar to that under the Clinton administration (Harrison 2010; Paterson 2009). But the rhetoric showed a desire to undermine the treaty, not just opt out. US rhetoric on climate change focused on the economic damage and potential job losses under the Kyoto Protocol. Interest groups representing coal and other climate-damaging industries were heavily invested in this narrative and presented a strong electoral disincentive for support for the Kyoto Protocol (Harrison 2010).

The strategy of skeptical engagement continued within the UNFCCC. While the United States continued to offer input on technical negotiations regarding the Protocol's markets or forests, it would refrain from any hint of discussions around new commitments, including walking out of some negotiation sessions. The "breakthrough in Bali" in 2007 was significant because the United States agreed to join the negotiations toward a new agreement. It was extremely reticent, going into the final plenary to either block consensus or not join it. Ultimately, the United States agreed after South Africa and other developing countries agreed that their emissions reduction efforts would be measured, reported, and verified (IISD 2007).

During this time, American foreign policy consistently advocated for a new approach to multilateral climate governance, using language and ideas from US-based think tanks and other advisors close to the Bush administration. Think tanks like the Brookings Institution and the Council on Foreign Relations put forward ideas palatable to the US administration. They billed the Kyoto Protocol as a collapsed, failed, and ineffective treaty before it entered into force. The aim was to advance ideas that may bring the United States back into global climate cooperation. As David Victor, writing for the Council on Foreign Relations, stressed: "By making it clear that the United States won't pretend to meet the Kyoto limits, the Bush administration has, for all intents and purposes, killed the Kyoto Protocol. Now it has a responsibility to build an alternative" (CFR 2003).

The targets in the Kyoto Protocol were often criticized. The president's Council of Economic Advisers chair argued that a "fixed emission limit harms the economy. Until we invent and commercialize new technologies to generate electricity and provide transportation … a fixed emission limit eventually means lowering economic growth" (Hubbard 2002). Similarly, the Brookings Institution argued that fixed targets reduced "emissions at any cost, rather than substantial emissions reductions at low cost" (McKibbin 2000). A low-cost emissions reduction plan would require a complete overhaul of the "flawed" CDM (Wilcoxen and McKibbin 1997, 2000).

Perhaps the most powerful rhetorical tool used by American think tanks was framing the Kyoto Protocol targets as a "top-down" infringement on national sovereignty. It helped create rhetorical space for the Bush administration to work on climate change on new terms. The Council on Foreign Relations identified three options for US involvement in multilateral climate action in the early 2000s as part of its Climate Policy Initiative. Two of the three options involve a successor agreement. One of these options for a new agreement is one of the first references to "top-down" and "bottom-up" treaty architectures. Invoking the success, and legitimacy, of the World Trade Organization's rules, CFR argued that bottom-up approaches to carbon markets could lead to successful treaties. A series of national or regional carbon markets could be interconnected, in a way similar to how currency markets had evolved (Victor 2004). Through this system, countries could all participate. The United States favored stronger national sovereignty and tried to diffuse the idea of a bottom-up approach.

The review of the Kyoto Protocol and the start of new negotiations in 2007 for a potential post-Kyoto agreement were also an opportunity for the United States. While the EU tried to use these openings to narrow the scope of discussions to the existing architecture by staying within the UNFCCC, the United States started alternative fora. It created alternative fora to advance its interests, widening the scope of discussions beyond the rules of the Protocol.

First, the United States formed the Asia-Pacific Partnership on Clean Development and Climate to entice these countries to set their own emissions reduction targets. The first six countries included China and India, representing half of global emissions at the time. At the inaugural meeting, the United States pledged USD 52 million to support the partnership in the first year (US Dept of State 2006). The Partnership was proposed as a complement to the Kyoto Protocol. In practice, the Partnership was a competitor to the Kyoto Protocol, largely because of its key features and timing, when the Protocol was yet to enter into force (McGee and Taplin 2006).

After the decision in Bali to start negotiations for a potential new agreement, the Bush administration started the Major Economies Process on Energy Security and Climate Change. Harlon Watson, the lead US climate negotiator, announced the Process at a UN meeting of the Convention Dialogue, explaining its primary aim to "seek agreement on the process by which the major economies would, by the end of 2008, agree upon a post-2012 framework that could include a long-term global goal, nationally defined mid-term goals and strategies, and sector-based approaches for improving energy security and reducing greenhouse gas emissions" (US Dept of State 2007). Seventeen countries, plus the EU, were invited.[2] The EU worried that the Process would weaken global climate cooperation (Bäckstrand and Elgström 2013). Perhaps due to EU efforts, the discussions only produced a declaration. The declaration referenced the value of a long-term greenhouse gases (GHG) goal. "Developed Major Economies" pledged to act "in the mid-term" to stabilize emissions and achieve absolute emission reductions, and "developing major economies" offered to act to deviate from business-as-usual emissions (White House 2008).

The United States repoliticized the Kyoto Protocol behaviorally and rhetorically. It first refused to ratify the Protocol, later establishing alternative fora to advance its preferences for global climate cooperation. These preferences were advanced in a way that would discredit the Protocol. It was a necessary rhetorical means to an end. Since President Bush had publicly denounced the Protocol, an alternative was necessary. The idea of a bottom-up treaty with nationally set targets helped create the distance between the Protocol and a new treaty.

14.2.3 Resolving the Impasse

Neither the repoliticization nor depoliticization strategies were proving effective in the run-up to the 2009 Copenhagen COP15 when a new agreement was to be adopted. The hope that others would follow its lead and support the Kyoto system had yet to be borne out for the EU. The United States had not realized any real

[2] Meeting participants: the US, the EU, France, Germany, Italy, United Kingdom, Japan, China, Canada, India, Brazil, South Korea, Mexico, Russia, Australia, Indonesia, and South Africa.

gains in the alternative fora it set up, leaving the new Obama administration with the options of working within the UNFCCC or walking away as his Republican predecessor had. The impasse was resolved by providing rhetorical wins for the United States, setting the stage for a new agreement while preserving the core institutions the EU heavily invested in.

The Copenhagen conference was largely a bust. Countries "took note" of the Copenhagen Accord, a short political document agreed to by a small set of leaders. But the Accord, and subsequent Cancun agreements, explicitly set down two important standards. First, developed countries would choose their own targets. Second, developing countries would choose their own actions subject to measurement, reporting, and verification. These standards raised the level of participation for developing countries while maximizing flexibility for developed countries.

The Accord also represented a compromise between the EU and the United States. It was the quintessential expression of a bottom-up agreement, starkly contrasting with the "top-down" Kyoto Protocol. This was the rhetorical win that the United States needed. It would not be backtracking to the unpopular Protocol but could move ahead to a new treaty based on national determination.

Yet this was not as sharp a break with the past as it seemed. As Depledge (2022) reminds us, Annex I countries had set their own Kyoto targets. The EU, in particular, set its target before the Kyoto Conference. It highlighted the difficult negotiations required to get agreement among the member states to reduce emissions "as a bubble" and refused to budge. The Accord (and later Paris Agreement) targets are not legally binding and have fewer rules governing their structure (e.g. a common baseline). But the nationally determined nature remains.

Since Copenhagen, the EU and the United States moved toward one another, building a bridge across the transatlantic divide (Kulovesi and Recio 2023; Paterson 2009). The EU began to pivot toward more of a facilitation role to try to get the United States and emerging economies on board (Bäckstrand and Elgström 2013). China, the United States, and the EU vied for leadership roles on the climate agenda.

14.3 Repoliticizing the Paris Agreement

With the Paris Agreement rules adopted in 2015, and superseding the Kyoto Protocol in 2020, it seemed that the climate regime was finally in "implementation mode." COPs from 2016 onward convened under themes calling for "action" or "implementation." Yet there is evidence of repoliticization, perhaps signaling a prolonged period of contestation rather than implementation.

The countries involved differ this time around. The United States, the EU, and other developed countries are repoliticizing Paris Agreement rules, largely in a way that isolates China, India, Saudi Arabia, and other larger developing countries that stand in defense of the Paris Agreement. The strategy is pocketing the Paris Agreement's wins, especially countries' wide participation. The United States and EU are now pushing for more. Burden-sharing emissions reductions remain a focal point, joined by calls to share the financial burden. In response, developing countries seek to add new institutions to fund loss and damage.

Strategies, however, remain largely the same. In repoliticizing mitigation rules, the United States and the EU have pushed for new language and concepts and initiated new fora. Attempts to introduce "major emitters" as a new category of countries have fallen flat, as have calls for new action "in this critical decade."

The coalition of Like-Minded Developing Countries (LMDCs)[3] has sought to depoliticize mitigation rules. The LMDCs would only accept a work program on mitigation, not new discussions that would change countries' varied responsibilities or the timelines for NDC submission. The 2022 mitigation work program decision limited it to a series of dialogues to conclude in 2026. There is limited scope to raise the issue to a political level or even get it on a negotiation agenda. Indeed, a proposal from the EU and Environmental Integrity Group to discuss the mitigation work program in the negotiations was unsuccessful (IISD 2023).

The scope of discussions on mitigation remains severely limited. Under the Paris Agreement, countries submit or update their NDCs every five years. There is no mitigation-related work at the UNFCCC in the intervening years, except for oversight of the market mechanisms. This new politicization conflict could augment these rules or distract from NDC implementation.

14.4 Conclusion

This chapter traces just one instance of a politicization conflict in the history of the UNFCCC. Shortly after the Kyoto Protocol was signed, it was already being questioned. Two competing strategies were at play, both behavioral and discursive, inside and outside the UNFCCC. The EU employed depoliticization efforts to shore up support for the Protocol. Markets were central for the EU to meet its mitigation targets and to build international networks and invest in the Kyoto Protocol's mechanisms. When opportunities arose, the EU sought more technical, roundtable discussions to bring the United States, China, India, and others into a Kyoto-like agreement. These efforts were successful in several ways. But, together, they show how much effort is involved in maintaining even a minimal level of stability, and the ease with which global rules can be reinterpreted or revised.

[3] Membership of the LMDC coalition has not been stable or public. Long-term core members include China, India, Saudi Arabia, Bolivia, and Cuba, among others.

Repoliticization efforts included refusal to ratify the treaty and creating alternate fora to build a support coalition for new ideas. New concepts recharacterized the national determination of mitigation targets. Instead of choosing a target and inscribing it in the treaty, as in the Protocol, the idea of "bottom-up" targets involved choosing a target that would remain non-legally binding. The rhetoric of the repoliticization efforts was adopted but not many of the actual recommendations. Carbon market reform was ultimately limited. The Paris Agreement maintains the primacy of national sovereignty but extends that flexibility to choose targets or other mitigation actions to all countries.

The resulting compromise may be more common in international politics than domestic arenas. A new government can hold a preponderance of power at the national level. States need to gain the support of others to achieve decisions by consensus. There is a fundamental need for compromise. Discursive strategies can be a powerful tool to achieve that consensus, by allowing countries to build common understandings. Yet this change can take considerably longer.

This Kyoto Protocol politicization conflict spanned roughly ten years and can reveal potential obstacles to the future implementation of the Paris Agreement. This recent round of politicization features similar tactics, but fault lines are more complex. Developed countries try to introduce new rules and concepts related to mitigation. Larger developing countries seek to limit the scope of this discussion. Developing countries are united in pushing for greater support to address the permanent effects of climate change through funding for loss and damage. Meanwhile, the Paris Agreement mechanisms are still in their infancy and already questioned.

For the Kyoto Protocol, countries on both sides found a way to accommodate one another, delaying climate action. The compromise resulted in a rhetoric of change that masked stability in mitigation rules. The central institutions of the Paris Agreement may stand, although the climate crisis is unlikely to withstand another delay.

References

Allan, J. I., and Bhandary, R. R. (2022). What's on the agenda? UN climate change negotiation agendas since 1995. *Climate Policy* 24(2), 153–163.

Bäckstrand, K., and Elgström, O. (2013). The EU's role in climate change negotiations: From leader to "leadiator." *Journal of European Public Policy* 20(10), 1369–1386.

Bäckstrand, K., and Söderbaum, F. (2018). Legitimation and delegitimation in global governance: Discursive, institutional, and behavioral practices. In J. Tallberg, K. Bäckstrand, and J. Dryzek, eds., *Legitimacy in Global Governance: Sources, Processes, and Consequences*. Oxford: Oxford University Press, pp. 101–118.

Betts, P. (2021). The EU's role in negotiating the Paris Agreement. In H. Jepsen, M. Lungren, K. Monnheim, and H. Walker, eds., *Negotiating the Paris Agreement: Insider Stories*. Cambridge: Cambridge University Press, pp. 111–138.

Black, I. and White, M. (2001). Summit admits to Kyoto failure. *The Guardian*, June 15. www.theguardian.com/environment/2001/jun/15/usnews.greenpolitics

CFR (Council on Foreign Relations). (2003). This summer's impending failure of the Kyoto Protocol. News release, January 8. www.cfr.org/news-releases/summers-impending-failure-kyoto-protocol

Christiansen, A., and Wettestad, J. (2003). The EU as a frontrunner on greenhouse gas emissions trading: How did it happen and will the EU succeed? *Climate Policy* 3(1), 3–18.

de Águeda Corneloup, I., and Mol, A. P. (2014). Small island developing states and international climate change negotiations: The power of moral "leadership." *International Environmental Agreements: Politics, Law and Economics* 14, 281–297.

Depledge, J. (2022). The "top-down" Kyoto Protocol? Exploring caricature and misrepresentation in literature on global climate change governance. *International Environmental Agreements: Politics, Law and Economics* 22(4), 673–692.

Eckersley, R. (2012). Moving forward in the climate negotiations: Multilateralism or minilateralism? *Global Environmental Politics* 12(2), 24–42.

EU. (2001). Presidency Conclusions, Göteborg European Council.

G8. (2005). Changement climatique: Eléments d'intervention de Jacques Chirac au rencontre et au déjeuner avec les Chefs d'Etats et le Gouvernement des pays émergents. Gleneagles Declarations, Working document, July 7. www.g8.utoronto.ca/francais/2005gleneagles/chirac-climat.html

Gampfer, R. (2016). Minilateralism or the UNFCCC? The political feasibility of climate clubs. *Global Environmental Politics* 16(3), 62–88.

Geoffrey, S., and Procter, G. (2005). G8 defies bombers as summit continues. *The Guardian*, July 7. www.theguardian.com/world/2005/jul/07/g8.uk

Gupta, J., and Ringius, L. (2001). The EU's climate leadership: Reconciling ambition and reality. *International Environmental Agreements* 1, 281–299.

Harrison, K. (2010). The United States as outlier: Economic and institutional challenges to US climate policy. In K. Harrison and L. M. Sundstrom, eds., *Global Commons, Domestic Decisions: The Comparative Politics of Climate Change*. Cambridge, MA: MIT Press, pp. 67–104.

Hubbard, R. G. (2002) Realism in cutting emissions, *New York Times*, February 15, p. A21.

Hughes, H., Vadrot, A., Allan, J. I. et al. (2021). Global environmental agreement-making: Upping the methodological and ethical stakes of studying negotiations. *Earth System Governance* 10, 100121.

IISD (International Institute for Sustainable Development). (2023). Summary and Analysis of the Bonn Climate Change Conference. Earth Negotiations Bulletin. https://enb.iisd.org/bonn-climate-change-conference-sbi58-sbsta58-summary

IISD (International Institute for Sustainable Development). (2007). Summary and Analysis of the Bali Climate Change Conference. Earth Negotiations Bulletin. https://enb.iisd.org/events/bali-climate-change-conference-december-2007/summary-report-3-15-december-2007

IISD (International Institute for Sustainable Development). (2005). Summary and Analysis of the Montreal Climate Change Conference. Earth Negotiations Bulletin. https://enb.iisd.org/events/unfccc-cop-11/summary-report-28-november-10-december-2005

IISD (International Institute for Sustainable Development). (2001). Summary and Analysis of the Marrakesh Climate Change Conference. Earth Negotiations Bulletin. https://enb.iisd.org/events/unfccc-cop-7/summary-report-20-october-10-november-2001

Kemp, L. (2016). Bypassing the "ratification straitjacket": Reviewing US legal participation in a climate agreement. *Climate Policy* 16(8), 1011–1028.

Kulovesi, K., and Recio, M. E. (2023). Fighting a hard battle with a soft weapon: Is international climate change law softening? In M. Eliantonio, E. Korkea-aho, and U. Mörth, eds., *Research Handbook on Soft Law*. Cheltenham: Edward Elgar, pp. 320–336.

Lisowski, M. (2002). Playing the two-level game: US President Bush's decision to repudiate the Kyoto Protocol. *Environmental Politics* 11(4), 101–119.

McGee, J., and Taplin, R. (2006). The Asia–Pacific partnership on clean development and climate: A complement or competitor to the Kyoto Protocol? *Global Change, Peace and Security* 18(3), 173–192.

McKibbin, W. J. (2000). Moving beyond Kyoto. Policy brief. Washington, DC: Brookings Institution. www.brookings.edu/articles/moving-beyond-kyoto/.

Paterson, M. (2009). Post-hegemonic climate politics? *The British Journal of Politics and International Relations* 11(1), 140–158.

Paterson, M., Tobin, P., and VanDeveer, S. D. (2022). Climate governance antagonisms: Policy stability and repoliticization. *Global Environmental Politics*, 22(2), 1–11.

Schattschneider, E. E. (1960). *The Semi-Sovereign People*. New York: Holt, Rinehart, & Winston.

Schreurs, M. A. (2004). The climate change divide: The European Union, the United States, and the future of the Kyoto Protocol. In N. J. Vig and M. G. Faure, eds., *Green Giants: Environmental Politics of the United States and European Union*. Cambridge, MA: MIT Press, pp. 207–230.

Szarka, J. (2012). The EU, the USA and the climate divide: Reappraising strategic choices. *European Political Science* 11(1), 31–40.

UNFCCC (United Nations Framework Convention on Climate Change). (2022). Decision on funding arrangements for responding to loss and damage associated with the adverse effects of climate change, including a focus on addressing loss and damage. Draft decision document no. FCCC/CP/2022/L.18, FCCC/PA/CMA/2022/L.20.

UNFCCC (United Nations Framework Convention on Climate Change). (2021). Glasgow Climate Pact. https://unfccc.int/sites/default/files/resource/cma2021_10_add1_adv.pdf

UNFCCC (United Nations Framework Convention on Climate Change). (2006). Review of the Kyoto Protocol pursuant to its Article 9: Submissions from parties. Document no. FCCC/KP/CMP/2006/MISC.3.

UNFCCC (United Nations Framework Convention on Climate Change). (2005). Decision on dialogue on long-term cooperative action to address climate change by enhancing implementation of the Convention. Document no. FCCC/CP/2005/L.4/Rev.1.

UNFCCC (United Nations Framework Convention on Climate Change). (2000). Tracing the origins of the Kyoto Protocol: An article-by-article history. Document no. FCCC/TP/2000/2.

UNFCCC (United Nations Framework Convention on Climate Change). (n.d.). Long-Term Work Programme on Finance. https://unfccc.int/topics/climate-finance/workstreams/long-term-climate-finance-ltf

US Dept of State. (2006). Asia-Pacific Partnership on clean development and climate: Concluding press conference. Sydney, January 12. https://2001-2009.state.gov/g/oes/rls/or/2006/59213.htm

US Dept of State. (2007). Building blocks for long-term cooperative action and their coherent integration into an appropriate and effective international response to climate change. Remarks to the Fourth Workshop: Dialogue on Long-Term Cooperative Action to Address Climate Change by Enhancing Implementation of the Convention, Vienna, August 27. https://2001-2009.state.gov/g/oes/rls/rm/2007/91483.htm

Victor, D. G. (2004). *Climate Change: Debating America's Policy Options*, Vol. 6. Council on Foreign Relations.

White House. (2008). Fact sheet: The Major Economies leaders meeting, July 9. https://georgewbush-whitehouse.archives.gov/news/releases/2008/07/20080709-6.html

Wilcoxen, P. J., and McKibbin, W. J. (1997). Salvaging the Kyoto climate change negotiations. Policy brief. Washington, DC: Brookings Institution. www.brookings.edu/articles/salvaging-the-kyoto-climate-change-negotiations/

Wilcoxen, P. J., and McKibbin, W. J. (2000). The next step for climate change policy. Research brief. Washington, DC: Brookings Institution. www.brookings.edu/articles/the-next-step-for-climate-change-policy/

15

US-China Relations and the Competitive Turn of Green Industrial Policymaking

JONAS NAHM

The past two decades have witnessed a resurgence of industrial policy – government interventions in the economy to produce outcomes in service of policy goals that markets would not yield on their own (Nahm 2022b). This resurgence has been evident in sectors related to decarbonization, where I refer to such policy interventions as "green industrial policy." Governments across developing and advanced industrialized economies have used such government interventions to combine climate and economic goals by trying to attract clean technology supply chains, including the manufacturing of wind turbines, solar panels, electric vehicles, and energy storage technologies.

Green industrial policies have long been deployed in the European Union and China, economies with statist policymaking traditions that were more amenable to government interventions in the economy to advance policy goals. But green industrial policies have also and increasingly been used in liberal market economies that long advocated laissez-faire approaches to governing the domestic economy. In the United States, one recent example is the Inflation Reduction Act (IRA) of 2022, which introduced a range of green industrial policy measures to accelerate the domestic development of clean energy industries, including using local content requirements. And the IRA is not the first US foray into green industrial policymaking. It was preceded by decades of tax breaks and regulatory incentives to help domestic clean energy sectors compete with fossil fuels (Nahm 2022b).

In this chapter, I examine the resurgence of green industrial policymaking, focusing in particular on the US response to China's dominance in clean energy sectors. I make three central points. First, I argue that the political logic of the energy transition makes industries related to decarbonization particularly prone to industrial policy interventions. Governments attempt to build political coalitions behind climate policy by promising new sources of growth and employment because the transition to the net zero economy requires vast public investments and subsidies (Nahm 2021). Second, I show that although governments in the United States and

Europe have long promised the development of domestic green industries due to such intervention, they have rarely been able to attract desired segments of global supply chains into domestic economies. Instead, clean energy sectors have generally relied on China for critical aspects of their manufacturing and raw materials production. Third, I make the case that this discrepancy between political promises and empirical results has fueled a politicization of industrial policy interventions. This has created more substantial incentives for firms to reshore production in critical industries, including in the United States. Such calls for reshoring likely understate the extent to which the world must continue to rely on Chinese manufacturing capacity for clean energy technology until alternative supply chains are established elsewhere. Perhaps more importantly, the increasingly protectionist justifications of industrial policy put such government measures in explicit tension with the need for global political collaboration to meet transnational decarbonization goals. Politicizing low-carbon industrial policies may create new economic coalitions behind climate policy if they indeed yield growing domestic supply chains and the associated jobs. The protectionist focus of such policies could also undermine the global supply chain arrangements required to decarbonize and threaten the basis for international political cooperation on climate goals. In contrast to existing literatures on climate politics that have correctly pointed to the importance of politicization to overcome entrenched corporate interests in fossil fuel industries, I suggest that politicization of green industrial policies can also have unintended consequences for the cross-national political and economic ties needed for long-term decarbonization (Paterson, Tobin, and VanDeveer 2022).

The chapter proceeds by highlighting the growing economic ambition behind climate policy, which is rooted in the political logic of promising growth in return for investments in decarbonization. Using the example of the United States I then show that such green industrial policies have in the past not prevented a growing reliance on China for the production of clean energy technologies. The final section of the chapter examines the dilemma resulting from this divergence of political promises and economic results. I suggest that governments have responded by inserting protectionist elements into green industrial policymaking, and conclude by reviewing the prospects of decarbonization under such a fragmented green industrial policy regime.

15.1 The Political Logic of Industrial Policy

Since the early 1990s, markets for low-carbon energy technologies have grown dramatically. According to the International Energy Agency, global investment in clean energy exceeded USD 1.4 trillion in 2022. Growing global spending on renewable energy technologies made up a significant fraction of such expenditures, but investments in grids, storage, and energy efficiency, and, more recently, rapidly accelerating sales of electric vehicles, also contributed to growing global markets for low-carbon technologies (IEA 2022).

Before rapidly falling costs for some clean energy technologies made them competitive with fossil fuels, growing markets for such technologies relied on government support, primarily through subsidies and regulatory incentives. Such measures included premiums paid to wind and solar energy generators, tax credits for purchasing electric vehicles and investments in energy efficiency, and a range of regulatory mechanisms to encourage utilities and car manufacturers to use and sell a growing share of low-carbon technologies. Due to the increasing expansion of clean energy markets and the significant fiscal investments and regulatory measures needed to maintain them, many governments were unsatisfied with being mere consumers of clean energy technologies. Instead, governments hoped that green industrial policies would allow them to build domestic industries that could invent, produce, and ideally export clean energy technologies in addition to deploying them domestically (Nahm 2021). In many economies, green industrial policies – including direct subsidies, tax breaks for manufacturers, local content requirements, and research and development (R&D) funding – were deployed to attract domestic clean energy supply chains.

A broader political strategy drove the connection between the adoption of clean energy technologies and the potential economic advantages in the form of industrial growth and local manufacturing activities. As part of government efforts to reduce greenhouse gas emissions, green industrial policies that offered the prospect of boosting employment and economic development were appealing to policymakers. These policies could grow political alliances supporting decarbonization beyond the core group of environmental advocates (Aklin and Mildenberger 2020). Such political support was significant for policies that entailed large public expenditures and threatened the business models of existing fossil fuel-based industries. But the promise of growth and employment also helped justify the additional financial burdens for private citizens from the energy transition, including for consumers of electricity asked to pay surcharges to offset the cost differential between traditional energy sources and higher-priced wind and solar technologies. Policies that pursued the dual objective of achieving emissions reductions while creating new sources of economic activity were easier to implement politically, not least because they expanded the political coalition in favor of decarbonization (Breetz, Mildenberger, and Stokes 2018; Meckling and Nahm 2021; Meckling et al. 2015). The growing use of green economic goals was a political tool to achieve policy stability in a climate policy arena historically shaped by conflict between powerful vested interests in the fossil fuel sector and a growing environmental movement.

The increasing use of green industrial policy led to a rise in state interventions in the economy, aimed at bolstering domestic clean energy supply chains. Governments combined established R&D support policies with subsidies to establish renewable energy markets, often specifically tied to local content regulations and other incentives to attract local industrial activity and manufacturing jobs. In the United States, policymakers believed that critical market failures for new technologies were primarily in basic R&D, where firms, if left to their own devices, would likely underinvest in these innovations. Thus, federal and state governments provided regulatory support and subsidies to create markets for these applications in sectors like clean energy, where new technologies were not yet competitive with the ones they intended to replace. The assumption was that public spending on technology development, combined with market support, would yield the emergence of new industrial sectors.

The desire to establish internal green energy supply chains became notably apparent in the United States following the worldwide economic crisis of 2009. The US government responded by implementing stimulus packages to provide unparalleled support to the local renewable energy industry. The 2009 American Recovery and Reinvestment Act (ARRA) was a critical piece of legislation in this effort, offering a variety of measures to facilitate this shift. This included a unique tax credit specifically aimed at clean energy production, along with loan assurances for wind and solar equipment producers. Furthermore, it funded training programs to prepare workers for the burgeoning clean energy sectors (Mundaca and Richter 2015). Government assets were purposefully utilized to boost development and promote economic restructuring by backing certain industrial fields. The green industrial policies launched during President Barack Obama's tenure were built upon widespread backing for the wind and solar industries at the state level. This support was prevalent as most states had enforced various renewable energy requirements. Such investment was frequently defended on the grounds that it would lead to job creation and economic expansion (Stokes and Warshaw 2017).

Other governments similarly intervened in energy markets to create new industrial sectors. The German government, for instance, explicitly justified continued support for wind and solar sectors with the ability of the domestic economy to capitalize on the "tremendous export market that will permanently secure growth and employment" (Bundesregierung 2005: 19). The Federal Ministry for the Environment projected in 2008 that, by 2020, green industries, encompassing renewable energy, recycling, and energy efficiency technologies, would outperform the German auto sector in terms of their impact on the country's gross domestic product (Bundesministerium für Umwelt 2008: 6).

China's public support for wind and solar energy was primarily driven by its industrial development objectives. The country viewed renewable energy sectors as potential export industries, aligning with its broader economic development strategy. This emphasis was evident in 2010 when renewable energy sectors were specifically listed as "strategic emerging industries" (SEIs). The SEIs sought to grow industrial sectors central to enhancing national competitiveness by providing preferential policy treatments, including low-interest loans, tax breaks, and R&D support. Under the direction of the central government in Beijing, companies were encouraged to reduce their reliance on international technology transfers. Instead, they were urged to focus on filling the remaining gaps in domestic supply chains, including the production of advanced manufacturing equipment essential for renewable energy technologies (US-China Business Council 2013). The 12th Five-Year Plan for the solar photovoltaic (PV) industry, part of China's policy of setting broad economic goals in five-year increments, called for 80 percent of production equipment to be manufactured domestically by 2015 (National Energy Administration 2011). The central government in Beijing aimed to capitalize on the emergence of new clean energy technologies as a strategic opportunity to secure a foothold in upcoming industries.

The growing promise of decarbonization technologies allowed governments to promise economic returns for investments in decarbonization, making it easier to use the vast public resources required to implement meaningful climate policy and potentially growing the political coalition of climate supporters.

15.2 Domestic Policy and the Emergence of Global Supply Chains

Despite government attempts to strategically position domestic economies in clean energy sectors, few countries were able to build wholly domestic supply chains for such technologies. Governments were able to help domestic firms enter some segments of global clean energy industries. Still, most countries remained locked in a division of labor in which major products and components had to be imported from abroad. China, in particular, came to dominate clean energy supply chains (Nahm 2021).

The contrast between government investments and the relatively constrained progress in developing domestic supply chains was especially notable in the United States, although similar patterns were observed in other regions worldwide. The United States held the position of being the largest investor in clean energy R&D and maintained a technological advantage in numerous critical areas. US companies continued to spearhead the advancement of cutting-edge technologies aimed at achieving more cost-effective and efficient decarbonization. These technologies included next-generation solar technologies, advanced battery chemistries, novel building materials, smart grid technologies, and sophisticated software for managing complex energy systems (Sivaram et al. 2020). Yet such new technologies needed to be commercialized and manufactured at scale, and little of that occurred domestically.

US policies focused on clean energy sectors under the assumption that market failures were present in upstream R&D, as well as downstream market demand creation. However, these policies did not place an equal emphasis on segments of the clean energy supply chains that lacked sufficient domestic support, particularly in manufacturing. The challenges in scaling up domestic production of clean energy technologies were indicative of long-term structural changes in the US economy. The decline of the manufacturing sector over time had raised concerns about the country's economic competitiveness and the ability of American firms to maintain their global leadership in developing new technologies without strong capabilities in commercialization and production on home soil. During the 1980s, the United States faced the fear of losing its competitive edge to Japan. Although Japan's subsequent economic crisis and the US IT boom in the 1990s diverted attention from domestic manufacturing issues, the underlying structural causes contributing to a diminishing share of high-wage manufacturing employment were never fully addressed (Berger 2013). China's entry into the World Trade Organization (WTO) in 2001 amplified import competition, yet the decline in the domestic manufacturing economy was not solely a result of this development. There were also other internal factors that contributed to this decline, and these factors preceded China's emergence as a major player in the global economy (Bonvillian 2017).

Despite significant investments, both public and private, in R&D that allowed the United States to maintain its leadership in pioneering innovative clean energy technologies, changes in the economy weakened the once strong connection between innovation and production. This link was crucial in translating technological breakthroughs into tangible industrial outputs. The structure of the domestic economy changed significantly over time, resulting in a considerable reduction in the number of manufacturing firms equipped with the capabilities to effectively scale up and produce new products and technologies. Over the years, multinational corporations in the United States increasingly turned to outsourcing and offshoring noncore production activities, as financial markets rewarded larger firms for focusing on their core competencies since the 1970s. The domestic manufacturing sector also encountered challenges in obtaining financing to move technologies from the research phase to full-scale production, particularly in emerging sectors like clean energy (Bonvillian 2017).

In clean energy markets, the United States had difficulty translating its position into tangible industrial outcomes, even though the widespread adoption of wind and solar technologies resulted in substantial service-sector industries for installation and maintenance. The United States achieved this position through effective policy interventions aimed at promoting the domestic deployment of clean energy technologies, leading to its emergence as one of the largest global markets for solar PV, wind power, and eventually electric vehicles. In cumulative solar installations, the United

States closely followed China and the European Union. Moreover, in 2021, it attained the third-largest market share for new solar PV installations. Similarly, in the wind industry, the United States positioned itself as the third-largest global market, ranking behind China and the European Union (IRENA 2021). At the same time, the United States became the third-largest market for electric vehicles (IEA 2021).

Nevertheless, the expansion of domestic markets for these technologies did not yield a parallel increase in domestic production capacity. Despite significant investments from the public sector in R&D and a commitment to supporting clean energy markets, sectors like wind, solar, and batteries in the United States continued to be reliant on imports for critical components. Among these sectors, the wind industry demonstrated the strongest domestic position, with the United States being one of five major global economies capable of producing all the essential components needed for modern wind turbines – nacelles, blades, towers, generators, and bearings. International and domestic turbine manufacturers set up production facilities within the United States, capitalizing on the advantage of proximity to end-user markets. Producing heavy components, such as nacelles, domestically resulted in substantial cost savings due to the expenses associated with transportation. However, even in the wind industry, the United States struggled to match the local content rates achieved by its European and Asian competitors. In 2019, wind turbine equipment imports into the United States accounted for a substantial USD 2.6 billion, with the largest portion comprising wind turbine blades and hubs from Brazil, China, and India. Turbine towers, which have long been a subject of trade conflicts with China, were primarily imported from India, Indonesia, and Vietnam. Estimates indicated that only about half of the manufacturing value of a wind turbine was generated domestically, which paled in comparison to Europe and China's local content rates of more than 90 percent for similar turbines (Bloomberg New Energy Finance 2021).

Until recently, the solar PV industry in the United States lacked the production capacity for key components such as ingots, wafers, or cells used in crystalline silicon solar PV modules. These modules represent the prevailing technology in the industry and are anticipated to remain crucial until 2035. As the cost of crystalline solar PV technologies significantly declined by 85 percent over the past decade, US companies that had focused on developing newer thin-film solar technologies faced challenges in penetrating the market (Hart 2020). Here, the United States possessed some domestic manufacturing capacity for producing thin-film modules, which comprised 16 percent of domestic solar demand but only accounted for 4 percent of global solar PV markets. One-third of thin-film modules installed in the United States were also manufactured there (US Department of Energy 2022b). The US solar industry supported a workforce of more than 230,000 individuals, with the majority employed in service-sector roles focusing on installation and maintenance. Despite nearly a decade of trade remedies concerning Chinese modules, these jobs relied heavily on imported technologies. Until recently, approximately 97 percent of the silicon wafers, the primary input for manufacturing solar PV cells, were produced in China. Moreover, Chinese firms operating in Vietnam, Malaysia, and Thailand manufactured silicon solar cells to produce solar PV modules intended for the US market. For the limited share of modules produced domestically, nearly two-thirds of the value was attributed to China due to the necessity of importing raw materials and essential components (US Department of Energy 2022b).

Similar to the solar industry, the battery sector also relied on technologies that were initially developed, at least in part, in the United States but have now shifted primarily to other regions. Aided by their own set of green industrial policies, Chinese battery manufacturers steadily improved their technological capabilities and expanded their dominance in production capacity, particularly for lithium-ion batteries extensively used in electric vehicles. China secured significant control over the production and refining of raw materials, essential for core material inputs. Furthermore, it bolstered its position as the major producer of subcomponents crucial for battery manufacturing. In comparison, the United States accounted for less than 1 percent of the global production capacity for cathode and anode materials, merely 3 percent for separators, and 7 percent for electrolyte production. China emerged as the dominant player with 63 percent of cathode manufacturing, 84 percent of anode materials production, 66 percent of separator production, and 69 percent of electrolyte production. Despite hosting 13 percent of the world's lithium-ion cell manufacturing capacity, largely attributed to Tesla's domestic investments, the United States remained highly reliant on imported subcomponents. In a remarkable feat, China's battery industry controlled approximately 80 percent of global lithium-ion cell production (US Department of Energy 2022a).

Although governments in the United States and Europe long promised the development of domestic green industries as a result of state interventions in the economy, they were not always able to attract the desired segments of global supply chains into domestic economies. Instead, clean energy sectors frequently relied on China for critical aspects of their manufacturing and raw materials production (Davidson et al. 2022; Nahm 2022a). The difficulty of building domestic clean energy industries threatened to undermine the key policy goal of achieving policy stability by expanding the political coalitions in support of decarbonization. High import quotas played into the hands of climate critics, who had long argued that climate policy is bad for the domestic economy.

15.3 The Politicization of Green Industrial Policy

China's rapid rise in clean energy sectors – and the large gaps in supply chains in Europe and North America – fueled a politicization of policies intended to promote the energy transition. Politicization in this context refers to the explicit use of policy tools that upend existing supply chain arrangements to favor domestic production. Industrial policies that long intended to forge the development of domestic clean energy sectors were now increasingly accompanied by trade conflict and other disputes with nations that were also trying to spur the development of domestic clean energy industries.

Governments and industry associations accused China of illegal subsidies and other policy practices that violated trade rules. More generally, the use of industrial policies to forge the development of domestic clean energy sectors by governments worldwide created considerable tensions in a global trading system ostensibly built around principles of equal treatment and nondiscrimination. The perhaps most controversial measure was the introduction of local content requirements, which, dating back to the early 2000s, were used in economies ranging from Canada to China to make eligibility for domestic subsidies conditional on local production. Local content requirements led to several trade disputes involving a range of different economies. The United States filed complaints against China and Japan, the European Union filed complaints against Canada, and reciprocal complaints were filed between the United States and India (Lewis 2014). Other interventions, such as domestic subsidies for manufacturing businesses and different types of financial support for domestic firms deemed illegal under WTO rules, also caused widespread disputes and led to a series of antidumping and countervailing duty measures to level the playing field for domestic producers (Meckling and Hughes 2017).

Simultaneously, trade policy took center stage as a pivotal instrument in green industrial policymaking, especially in the United States. The implementation of trade policy arose as a reaction to China's increasing control over supply chains for clean energy technologies, including solar, wind, and battery storage. The US government initiated trade measures on several Chinese imports to offset China's prevailing dominance, commencing in the mid 2000s. The dynamics of trade within the clean energy industry highlighted conflicting interests between firms seeking domestic manufacturing and a service sector prioritizing installation and maintenance, which continued to heavily rely on Chinese imports (Meckling and Hughes 2017; Nahm 2022b).

A noteworthy case in 2011 involved a group of US wind turbine tower manufacturers filing a trade grievance against Chinese tower companies. The US International Trade Commission subsequently approved antidumping tariffs in 2013. While US-based tower manufacturers welcomed this move, those in the installation and maintenance sector expressed concerns that these trade remedies might lead to higher prices and hinder the deployment of clean energy technologies. Similarly, in 2010, a coalition of US solar manufacturers successfully pursued trade remedies against Chinese solar panels. However, the "Coalition for

Affordable Solar Energy" failed in its attempt to prevent the implementation of these remedies, which were eventually enforced in 2012. In response, Chinese manufacturers adjusted their supply chains to include Malaysia and Taiwan, prompting the US Department of Commerce to expand the scope and coverage of the tariffs. Consequently, the Solar Energy Industries Association publicly aligned with installers in opposition to manufacturers (Nahm 2021). Despite facing opposition, the imposition of trade remedies was deemed indispensable and persisted during the Trump administration. These measures were designed to address Chinese subsidies and the adverse effects on US solar cell manufacturers caused by artificially low-priced Chinese solar cells. However, domestic solar installation and maintenance companies expressed concerns that these remedies might lead to higher prices and subsequently reduce demand. Significantly, the impact of these trade remedies on downstream installation and maintenance sectors was not originally included within the investigative scope of the US Department of Commerce and the US International Trade Commission (Groom 2019).

In 2021, the Biden administration subsequently reviewed critical industrial sectors' domestic supply chains to scrutinize the dependence on Chinese inputs for essential technologies. The review aimed to evaluate the economic and security implications of such reliance (Igogo 2022). The tariffs imposed on Chinese clean energy goods due to prior trade investigations remained in effect during the Biden administration. They were intended to stimulate the relocation or broadening of supply sources beyond China. Nonetheless, other measures that would enhance domestic companies' competitiveness, such as public investments in vocational training and improved financing for demonstration, commercialization, and manufacturing, were not implemented.

In conjunction with trade policy, various other measures were taken to actively facilitate the reshoring of manufacturing from China. Under the Biden administration, the Defense Production Act (DPA) played a significant role in addressing domestic supply chain risks linked to national defense and security concerns. In particular, the DPA was instrumental in expediting the domestic production of five critical clean energy technologies, such as solar, where the United States had become overly reliant on global supply chains, especially China's dominant role. Through the DPA, the US government provided minimum order guarantees, extended loans to manufacturers, and established domestic content standards, all aimed at bolstering domestic manufacturing capacity (Sherlock et al. 2022).

In conjunction with policies aimed at addressing China's dominant position in clean energy supply chains, several initiatives were implemented that linked support for the energy transition to stringent local content standards. Notably, the IRA, passed in August 2022, played a key role in accelerating the adoption of electric vehicles (EVs) and clean energy within the US power grid. The IRA offered generous incentives for local production, including a USD 7,500 tax credit for new EVs and a USD 4,000 tax credit for used ones. A significant departure from previous practices, eligibility for these tax credits was now linked to the location of technology manufacturing, with a clear objective of greatly expanding domestic clean energy manufacturing. To be eligible for the full EV tax credit, a minimum of 40 percent of the materials used in the battery had to be sourced and processed in the United States or a country that holds a free trade agreement with the United States (Sherlock et al. 2022).

Additionally, the battery and its components had to be manufactured in North America. Similarly, the IRA provided bonus credits for clean energy technologies manufactured in the United States for investments in and generation of zero-emissions electricity. Tax credits for wind and solar installations, which have long been a core element of US government support for domestic clean energy markets, were extended under the act and now also included local content requirements to encourage the reshoring of clean energy supply chains (Sherlock et al. 2022).

The shift in US policy toward using local content requirements and measures previously opposed by the United States at the WTO sparked competitive efforts to nationalize clean energy sectors in other countries. Canada and the European Union responded with their own industrial policy measures, injecting new capital into clean energy sectors while undermining existing supply chain arrangements that have led to declining wind, solar, and battery technologies costs (McNamara 2023). This marked a departure from past decades; governments now aggressively intervened in global supply chains, using China's dominance in global clean energy sectors to justify a departure from global trade arrangements and growing direct investments in domestic industries.

As governments renewed their efforts to attract clean energy supply chains through trade policy and government spending on domestic firms, they did not always address the institutional reasons that in the past stymied attempts to draw the manufacturing of technologies required for decarbonization. Domestic institutions central to building clean energy economies include supportive financial systems willing to lend to manufacturers, vocational training institutions capable of meeting the workforce needs of the industries governments are trying to build, and R&D organizations supporting (existing) firms entering clean energy sectors, among others. These institutions can complement industrial policies in meeting their objectives. Yet the current resurgence of green industrial policy was not matched with equally ambitious reforms of the institutions governing the domestic economy. The IRA in the United States prioritized the revival of domestic manufacturing. However, it fell short of addressing some of the fundamental structural challenges that in the past have hindered the transformation of public investments in R&D and domestic markets into a thriving and competitive clean energy manufacturing sector.

The competitive turn of green industrial policymaking highlights a central tension between the need to build broader political coalitions behind climate policy to achieve policy stability and the destabilizing effect of undermining existing trade and political relationships. Large investments in domestic clean energy sectors were now becoming increasingly conditional on production locations and supply chain requirements, yet risked alienating trade partners who would eventually have to cooperate on global decarbonization goals. At the same time, the growing use of trade policy and local content requirements threatened to potentially increase the cost and slow the pace of decarbonization, as it entailed replacing and supplementing existing supply chain arrangements while prioritizing resilience over efficiency.

15.4 Conclusion

The resurgence of green industrial policies could potentially upend existing politics of climate change by creating economic and industrial interests that are invested in and are benefiting from decarbonization. Instead of trying to buy off industries invested in fossil fuels, green industrial policies could grow the political coalition in favor of continued decarbonization, thereby shifting the balance of economic interests away from the status quo (Meckling and Nahm 2021). The focus on domestic supply chains in this new generation of green industrial policy follows a political logic: Creating jobs in clean energy sectors – through local content requirements and trade policy – can help build new coalitions behind climate policy, including in areas where climate change has not yet been a priority of voters. In the United States, the IRA is a new chapter not just for US climate policy but also for rethinking the role of the state in taking advantage of economic opportunities in rapidly growing global clean energy sectors. The bill is just the starting point of a much broader industrial transformation for the United States and other economies that have started to emulate the US initiative. Other Western economies have responded by investing in their own clean energy industries to ensure economic gains accrue domestically.

Economists have long warned that industrial policies are prone to capture, can lead to inefficiencies and the misallocation of capital, and trigger rent-seeking behavior on the part of the private sector (Hallegatte, Fay, and Vogt-Schilb 2013). Yet the challenges of green industrial policy may ultimately be less about governments picking the wrong winners than about getting right the domestic and international politics of industrial policymaking. As I have outlined in this chapter, the political logic of industrial policy leads governments to overpromise on industrial policy outcomes in sectors related to decarbonization. Such policies themselves now often directly conflict with the supply chains that have made clean energy technologies cheap and accessible. Trade policies can help the growth of domestic industries but can also lead to retaliation from other countries such as China, which continue to control key inputs for clean energy supply chains.

Politicizing low-carbon industrial policies could be a political tool to achieve policy stability by creating new economic coalitions behind climate policy through growing domestic supply chains and the associated jobs. But the approach also risks undermining the global economic order required to rapidly decarbonize and threatens the basis for long-term international political cooperation on climate goals. The stakes of balancing politicization and stability have never been higher. Because climate and economic outcomes are now so closely linked in countries that have made support for decarbonization conditional on industrial development outcomes, failing to meet industrial policy goals will jeopardize climate targets as well.

References

Aklin, M. and Mildenberger, M. (2020). Prisoners of the wrong dilemma: Why distributive conflict, not collective action, characterizes the politics of climate change. *Global Environmental Politics* 20(4), 4–27.

Berger, S. (2013). *Making in America: From Innovation to Market*. Cambridge, MA: MIT Press.

Bloomberg New Energy Finance. (2021). *Wind Trade and Manufacturing: A Deep Dive*. Report, February.

Bonvillian, W. B. (2017). Advanced manufacturing: A new policy challenge. *Annals of Science and Technology Policy* 1(1), 1–131. https://doi.org/10.1561/110.00000001

Breetz, H., Mildenberger, M. and Stokes, L. (2018). The political logics of clean energy transitions. *Business and Politics* 20(4), 492–522.

Bundesministerium für Umwelt (2008). Konsolidierte Fassung der Begründung zu dem Gesetz zur Foerderung Erneuerbarer Energien im Waermebereich, BMU. Berlin.

Bundesregierung (2005). Wegweiser Nachhaltigkeit 2005 – Bilanz und Perspektiven: Kabinettsbeschluss vom 10. August 2005. Berlin: Presse- und Informationsamt der Bundesregierung.

Davidson, M. R., Karplus, V. J., Lewis, J. I., Nahm, J. and Wang, A. (2022). Risks of decoupling from China on low-carbon technologies. *Science* 377(6612), 1266–1269.

Groom, N. (2019). US solar group says Trump tariffs killing jobs; White House says "fake news." *Reuters*, December 4. www.reuters.com/article/uk-usa-solar-tariffs/u-s-solar-industry-to-lose-62000-jobs-due-to-trump-tariffs-study-idUSKBN1Y71V8

Hallegatte, S., Fay, M. and Vogt-Schilb, A. (2013). Green industrial policies: When and how. Policy Research Working Paper No. 6677, World Bank, Washington, DC. https://doi.org/10.1596/1813-9450-6677

Hart, D. (2020). *The Impact of China's Production Surge on Innovation in the Global Solar Photovoltaics Industry*. Washington, DC: Information Technology & Innovation Foundation.

IEA (International Energy Agency). (2021). *Global EV Outlook 2021*. Paris: IEA.

IEA (International Energy Agency). (2022). *World Energy Investment 2022*. Paris: IEA. https://iea.blob.core.windows.net/assets/b0beda65-8a1d-46ae-87a2-f95947ec2714/WorldEnergyInvestment2022.pdf

Igogo, T. (2022). *America's Strategy to Secure the Supply Chain for a Robust Clean Energy Transition*. U.S. Department of Energy Response to Executive Order 14017. https://doi.org/10.2172/1871491

IRENA (International Renewable Energy Agency). (2021). *Renewable Capacity Statistics 2021*. Abu Dhabi: IRENA.

Lewis, J. I. (2014). The rise of renewable energy protectionism: Emerging trade conflicts and implications for low carbon development. *Global Environmental Politics* 14(4), 10–35.

McNamara, K. R. (2023). Transforming Europe? The EU's industrial policy and geopolitical turn. *Journal of European Public Policy*, 1–26. https://doi.org/10.1080/13501763.2023.2230247

Meckling, J. and Hughes, L. (2017). Globalizing solar: Global supply chains and trade preferences. *International Studies Quarterly* 61(2), 225–235.

Meckling, J., Kelsey, N., Biber, E. and Zysman, J. (2015). Winning coalitions for climate policy. *Science* 349(6253), 1170–1171. https://doi.org/10.1126/science.aab1336

Meckling, J. and Nahm, J. (2021). Strategic state capacity: How states counter opposition to climate policy. *Comparative Political Studies* 55(4), 493–523. https://doi.org/10.1177/00104140211024308

Mundaca, L. and Richter, J. L. (2015). Assessing "green energy economy" stimulus packages: Evidence from the US programs targeting renewable energy. *Renewable and Sustainable Energy Reviews* 42, 1174–1186.

Nahm, J. (2021). *Collaborative Advantage: Forging Green Industries in the New Global Economy*. New York: Oxford University Press.

Nahm, J. (2022a). Green growth models. In L. Baccaro, M. Blyth, and J. Pontusson, eds., *Diminishing Returns: The New Politics of Growth and Stagnation*. New York: Oxford University Press, pp. 443–464.

Nahm, J. (2022b). *Reimagine: Clean Energy Technology and US Industrial Policy*. Center for New American Security, Report, September 7. www.cnas.org/publications/reports/reimagine-clean-energy-technology-and-u-s-industrial-policy

National Energy Administration (2011). 12th Five-Year Plan on Solar Power Development [国家能源局文件, 国能新能(2012)194号]. Beijing: National Energy Administration.

Paterson, M., Tobin, P. and VanDeveer, S. D. (2022). Climate governance antagonisms: Policy stability and repoliticization. *Global Environmental Politics* 22(2), 1–11. https://doi.org/10.1162/glep_a_00647

Sherlock, M. F. et al. (2022). *Tax Provisions in the Inflation Reduction Act of 2022 (H.R. 5376)*. Congressional Research Service, Report. https://crsreports.congress.gov/product/pdf/R/R47202

Sivaram, V., Cunliff, C., Hart, D., Friedmann, J., and Sandalow, D. (2020). *Energizing America: A Roadmap to Launch a National Energy Innovation Mission*. New York: Columbia University SIPA Center on Global Energy Policy.

Stokes, L. C. and Warshaw, C. (2017). Renewable energy policy design and framing influence public support in the United States. *Nature Energy* 2, 17107. https://doi.org/10.1038/nenergy.2017.107

US-China Business Council (2013). *China's Strategic Emerging Industries: Policy, Implementation, Challenges, & Recommendations*. Washington, DC.

US Department of Energy (2022a). *Grid Energy Storage: Supply Chain Deep Dive Assessment*. US Department of Energy Response to Executive Order 14017, "America's Supply Chains." www.energy.gov/sites/default/files/2022-02/Energy%20Storage%20Supply%20Chain%20Report%20-%20final.pdf

US Department of Energy (2022b). *Solar Photovoltaics: Supply Chain Deep Dive Assessment*. U.S. Department of Energy Response to Executive Order 14017, "America's Supply Chains." www.energy.gov/sites/default/files/2024-12/Solar%2520Energy%2520Supply%2520Chain%2520Report%2520-%2520Final%5B1%5D.pdf

Part V

Reflections

16

The Politics of Stability and Politicization of Change

The Carbon Trap and Just Transition

STEVEN BERNSTEIN AND MATTHEW HOFFMANN

Stability per se is not a problem. It is the substance of what is locked in – the climate damage, the unjust/unequal hierarchies, the extractive system – that is the problem (see also Paterson, Tobin, and VanDeveer, Chapter 1, this volume; VanDeveer, Chapter 5, this volume). Certainly, disruption of the stable dynamics that comprise the current system of carbon lock-in (Unruh 2000) is desirable. But if a policy environment was stable that promoted and reinforced just low-carbon systems and societies, stability would be good.

The premise of this volume is that desirable disruption – unless driven by exogenous shocks – is constituted by politicization. The editors of this volume propose that even apparently desirable stability in "long-term emissions reductions pathways" or "policy design over time" – the second and third forms of stability they identify – if without politicization, will lead to ruin. For example, they worry that even stability in "long-term emissions reduction pathways" can entrench policy designs that in practice will lead to "scope expansion," that introduce "new sources of emissions or sinks into the policy and political debate." More broadly, they write that academics who promote policy stability can get "co-opted by the policymakers" who have a "deeper" commitment to stability than decarbonization and that prioritizing stability is "naïve" or "designed to avoid conflict" with entrenched interests, which has the effect of stabilizing policy regimes those interests support. The corollary is that stability is constituted by depoliticization in practice even when well-intentioned policy actors and analysts support stable policies they believe are desirable. We challenge this dichotomy, taking up the editors' challenge to add nuance to how forms of stability and politicization interact. We do so by focusing on the *political project* of achieving *both* desirable disruption *and* desirable stability.

Our central argument is that the desired relationship between policy stability and politicization changes depending on the structural and institutional conditions in place that reinforce or support the transformation of carbon lock-in (Bernstein and Hoffmann 2019). We introduce what elsewhere we have called the "carbon trap" as a way to contextualize changes in this relationship (Bernstein and Hoffmann 2019). The carbon trap heuristic identifies thresholds below which systems are "locked in" to carbon, prone to getting stuck in climate policy trajectories, or entrench "solutions" that prevent desirable transformative action. Below those thresholds, stability is undesirable and the relationship between stability and politicization is more dichotomous. Stability under these conditions tends to get stuck in an "improvement trap" where even policies that initially reduce emissions can build political and economic resistance to further change or are removed from political arenas (i.e. depoliticized).[1] The "trap" is that below this threshold, such policies are inadequate or contrary to required sociopolitical transformations. Above the threshold, the relationship between stability and politicization can change and be more synergistic, where the former reinforces political coalitions, norms, and capacities for transformative trajectories.

We proceed in three steps. First, we further unpack and reframe the false dichotomy between stability and politicization to show that both involve politics. Second, we introduce the carbon trap as a way to contextualize the relationship and identify desirable and undesirable stability. Third, we explore whether and how the concept and politics of just transition offer ways to understand and pursue desirable politicized disruption and catalyze stable policy and systems that enable it.

16.1 The False Dichotomy of Politicization and Stability

It is a category mistake to see stable policies as depoliticized. Attention to politics is central. As Paterson, Tobin, and VanDeveer (2022: 4) argue, "[i]t is an empirical mistake to assume that climate policy can be addressed 'outside politics'. Politics – when understood as conflicts of interest and power, and the ongoing necessity of collective decision-making – is simply intrinsic to social life."

Building coalitions and designing institutions that shape and shove politics or entrench new path dependencies entails struggle and contestation over normative change, often with incumbent interests, even when the goal is achieving a new *stable* status quo. The practical concern for climate action, however, is that too many initiatives or new institutional arrangements touted as "solutions" are not fully disruptive of carbon lock-in. The search for policy stability is thus potentially dangerous when the substance of what is stable, and thus reinforced, is insufficient to prevent or counteract the climate crisis. Especially problematic is policy stability as an end. If the policy interventions that will plausibly achieve stability now, during carbon lock-in, are unlikely to disrupt the status quo, then stability as a

[1] For the implications of removal, see Bauer and Knill's (2014: 39) discussion of "policy dismantling by arena-shifting."

goal is misguided. Worse, it may "lock in" stable but insufficient climate policies. Under such conditions, "repoliticization" of these policies or institutions – bringing them back into political arenas that open up contestation and expose power relations – is desirable.

Similarly, as scholars or policy actors, we may be tempted to identify institutional or structural conditions that lock in desired goals of just decarbonization policy trajectories toward low-carbon societies. However, doing so frequently neglects attention to the political conditions and processes required to create, maintain, and reinforce those institutions and structures. It risks producing naïve "solutions" that assume away the political struggles and processes that shape or undermine the conditions necessary for desirable stability. Normatively, we may also overestimate our ability to identify the requirements of just transformations and undervalue the importance of politicization and disruption in defining and motivating practices of just transition required to address the climate crisis.

These practices of climate action have structural and institutional consequences. In fact, that is the aim of climate action – replacing one status quo with a hopefully better one. If successful, transforming structures and institutions can tilt the system in ways that mitigate the size or strength of counter-coalitions and strategic action of incumbents or other opposing forces that aim to undermine or reverse specific policies or broader transformations. For example, there is a rich literature on institutional design, path dependency, and policy feedback that focuses on how institutions that reinforce political coalitions in favor of desirable policies, support normative change, or create material or ideational (e.g. reputational, social, or learning) incentives for ongoing policy responses in a particular direction are important for those policies to maintain or increase strength and durability (Auld et al. 2021; Levin et al. 2012; Lockwood 2022; Meckling 2019; Millar et al. 2021).

Institutions here do not replace politicization – their mere existence does not remove an item from contestation or debate (Kuzemko 2016). Analytically, this understanding of institutions recognizes that politics is not only about interest and ideas constantly battling it out; it is also about how they are shaped and channeled, recognizing the politics of design and institutional effects. Ignoring those politics gives ground analytically to existing institutional arrangements, as if they are immutable structures as opposed to themselves being the products of political struggles, power, and ideas that have become entrenched and reinforced over time.

Implicit in these debates is the volume editors' concern that overly focusing on locking in policy trajectories by attempting to create legislative or institutional firewalls will ultimately fail because either (1) it underestimates opposing or incumbent forces, especially of well-documented corporate interests and lobbying (Allan 2020; Hochstetler 2020; Levy and Newell 2005; Stokes 2020); (2) inadvertently produces stasis or undermining feedbacks (Millar et al. 2021; Patterson 2023); or

(3) limits political struggle and motivation needed for deeper transformation and system change. We are agnostic over whether these effects are as pernicious as some of the existing literature suggests but believe an analytic position that examines how such arrangements create *political* conditions for ongoing policy toward desirable ends counters the claim of such moves being depoliticizing.

Normatively and pragmatically, our argument falls into the camp of those who see the merits of long-term policy trajectories. "Stability," however, is a poor descriptor for such trajectories if it implies simple continuation, unchanging, conservative, or constant. Rather, our own work and that of others interested in entrenching, scaling, and reinforcing transformative policy trajectories is owing to a concern in building and maintaining support for transformative pathways that resist or counter tendencies of backsliding or outright oppositional coalitions and forces (Jordan and Moore 2020; Millar et al. 2021; Rosenbloom et al. 2019). This reflects our normative commitment to transformation that informs virtually all the work under discussion in this volume.

Pragmatically, transformative pathways lock in trajectories in the sense that, by definition, they must continually support and reinforce change in the direction of the transformative goal. However, they need not lock in specific institutions, policy choices, or instruments. That would be counterproductive for reasons well-articulated by those in the "politicization" camp but also by institutional scholars. Empirical work on policy feedback, for example, has found that even positive feedback mechanisms on some aspects of policy – for example a particular policy instrument – can have unintended consequences for other levels. Moore and Jordan (2020) found this exact pattern in their study of the EU's emissions trading system, where positive feedback led to the strengthening of the policy instrument (the trading scheme) but weakened support for the broader objective of reducing overall emissions.

If institutions and instruments lead to unintended or socially undesirable outcomes, politicization not only "should" occur but inevitably will occur. Systems ought to be open and encourage politicization *in order to* maintain trajectories. Here, the arguments of Auld et al. (2021) about building support and entrenching policy goals while fostering politicization over policy means are essential.

Focusing on the stability of some goals but politicization and opportunities for learning about all manner of means, participation, and accountability (see Auld et al. 2021) recognizes the political work that the editors, and scholars such as Jordan and Moore (2020), highlight as necessary for securing just transition trajectories. Conversely, permanent revolution is unlikely to produce the necessary conditions for actors to follow through with commitments, engage in long-term planning, encourage investments, or reduce uncertainty in ways that build confidence in a transformative trajectory. Nor will it help protect against reactionary reversals or prevent the waves of apparent progress followed by backsliding that have characterized many climate policies and transformative efforts observed in many countries.

These dilemmas go to the heart of the stability/politicization friction. We need disruptive politics to challenge and unsettle the status quo *and* we need that to happen in ways that catalyze policy stability and even lock in desirable substantive policy and systems that support just decarbonization – the political goal. As scholars, we need an analytic and normative way forward that places politics at the center of analysis as opposed to ignoring or silencing it. In sum, we need a framework that neither depoliticizes stability nor fetishizes disruption.

Addressing this tension and overcoming the pattern of some progress and getting stuck or backsliding has been the central focus in our work on the "carbon trap" (Bernstein and Hoffmann 2019). To overcome this trap, politicization is exactly what comes into focus *to produce* stable trajectories toward decarbonization. A key dynamic the trap illustrates is that below a certain threshold overt politicization and disruption of the status quo is needed. As the old politics of the status quo is undermined and new institutional and structural conditions support a politics that shifts momentum toward just decarbonization, the emphasis shifts to a politics of stability.

16.2 The Carbon Trap

One reason that policy stability is such a source of handwringing and analysis is that climate policies in many places have failed to be durable (Jordan and Matt 2014; Jordan and Moore 2020; Levin et al. 2012).[2] Policy interventions (carbon pricing, renewable energy portfolios, infrastructure spending, subsidies for low carbon products, etc.) frequently manage to get started (see Bernstein and Hoffmann 2018; Bulkeley et al. 2014; Hoffmann 2011) but often get stuck and are even reversed (Breetz et al. 2018; Meckling 2019; Rabe 2018; Stokes 2013, 2020). This pattern is observable whether states, non-state, or corporate actors initiate and govern the policies or actions, and whether they are domestic or transnational. That may not be surprising in a world locked in to the use of fossil energy. Yet fully grasping the source of the *failure* of policy stability is crucial for navigating the stability/politicization divide at the heart of this volume.

Figure 16.1 shows the logic of the carbon trap graphically. The x-axis is the current level of decarbonization in a system, and the y-axis is the expectation about the future level of decarbonization given the current state. The flat parts of the trajectory (we label them carbon lock-in and decarbonization) represent relatively stable state spaces where the array of factors (political, economic, technological,

[2] This section and next one draw significantly on and reproduce parts of Bernstein and Hoffmann (2019).

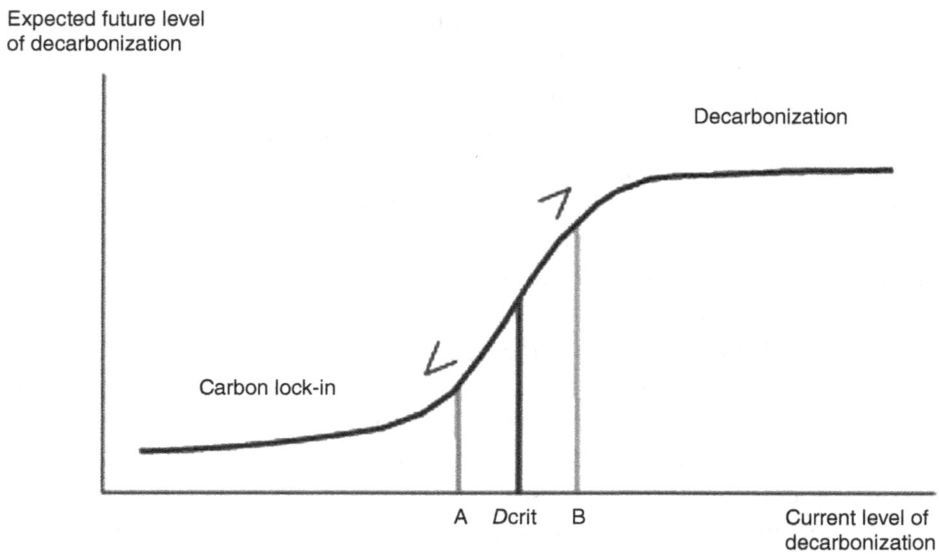

Figure 16.1 The carbon trap.

cultural) are aligned to generate reinforcing dynamics. The arrows represent the direction of expectations. If a system (city, market, nation-state) is currently below (point A) the critical threshold (*D*crit), the expectation is that the future state of the system will move toward the carbon lock-in attractor, that is, a label we borrow from complex system analysis of a relatively stable state space where the array of relevant factors (political, economic, technological, cultural) are aligned and generate reinforcing dynamics.

When a new climate policy is introduced, it has the potential to move the system along the trajectory toward greater decarbonization. The *D*crit threshold could potentially be breached either by a sudden shift such as a technological breakthrough or through the progressive accumulation of incremental changes over time. Regardless, unless a new policy helps the targeted system to breach the *D*crit threshold (B), political-economy forces in the system will tend to drive the system back toward the carbon lock-in attractor. These forces include technology and technological practices (the range of choices available and the pace of innovation); economic factors (sunk costs, investment cycles, energy markets, cost structures, risk calculations of holders of climate-vulnerable or climate-friendly assets, and so on); cultural inertia (what people want and the practices that they consider normal); and political dynamics (coalitions, norms, institutions, and interests – including those that affect calculations of risks and benefits – that promote or resist change).

Below the critical threshold, the carbon locked-in status quo is dominant and needs to be disrupted (politicized) because transformative policies and action will be unstable. Policy interventions must ultimately navigate counter-coalitions supported by incumbent interests and industries, campaigns that appeal to entrenched cultural norms and practices, and extant institutional arrangements that often favor existing policies in path-dependent ways that can make change difficult to sustain. These forces are not static. Interventions engage and generate politics that can create positive and negative feedback, mobilization, and backlash (e.g. Breetz et al. 2018; Jordan and Matt 2014; Patterson 2023).

Consider a municipal policy to install electric charging stations to increase decarbonization in its transportation system. This policy would move the city away from the lock-in attractor and up the decarbonization curve a modest amount (perhaps to point A in Figure 16.1). However, once in place, other political-economic factors in the city and beyond could drive the city's transportation system back toward the carbon lock-in attractor if the D_{crit} threshold is not breached. These threshold-determining factors could include availability of affordable electric vehicles (EVs) determined by the actions of incumbent industries and government policies and investment decisions; cost and accessibility of charging station technology and places to put them within and outside the city, as well as battery capacity, which are intertwined technological, policy, and even psychological (range anxiety) issues; cachet or stigma of driving an EV, which could have both demographic and geographical determinants; and complementary incentives or disincentives at the sub-state or national level for purchasing EVs, which depend on interests and coalitions at other levels of government.

Policies designed to move systems toward decarbonization do not always make it over the threshold. Apropos of this hypothetical example, in 2019 in Ontario, Canada, the provincially controlled regional public transportation agency removed EV charging stations from its parking lots, citing high costs and low demand after an election shifted the political conditions for climate policy (Boisvert 2019). That same government ended a rebate program for EVs (Benzie 2018). Not surprisingly, the result in the case of EVs is that, in 2023, Ontario sales continued to lag provinces British Columbia and Quebec, which have rebates and greater infrastructure investment (Rabson 2023). The larger global pattern of EV sales shows similar patterns. On the one hand, many major auto companies worldwide appear to have reached a tipping point in their plans to transition to EV production, and the market share of EVs continues to rise, given a combination of technological change, costs and subsidies, anticipated and real regulation, and normative change. Yet consumer surveys and purchasing patterns show that in many places sales are lagging expectations precisely because of inadequate infrastructure and range anxiety that give consumers pause. Meanwhile, investors and analysts are reinforcing negative market reaction over fears of oversupply and weaker than expected consumer demand (Mullaney 2023). Thus, even when parts of the system shift, overcoming the threshold where the decarbonization attractor takes over requires multiple changes, many of which require ongoing politicization to overcome. Examples like these, across many sectors, are, unfortunately, common, for reasons ranging from changes in government to unexpected price shifts to active opposition (Millar et al. 2021; Stokes 2013).

The key implication of the carbon trap for the politicization/stability divide is that if a system is located in the carbon lock-in equilibrium, we should not expect climate policies to be stable. We should expect that they face challenges and struggle to endure because the attractor is carbon lock-in. From this perspective, it is not surprising that a focus on policy stability has arisen. Those that care about climate change observe that many attempts to disrupt carbon lock-in and the status quo fail or get retrenched, and we are running out of time. A strong desire for durability and constraining our future selves is an understandable response (Levin et al. 2012). An analysis of the carbon trap adds to the debate on stability a focus on the political dynamics that stability generates. Climate interventions must disrupt coalitions and the normative/institutional context that supports the status quo to breach the critical threshold, while also generating the stability necessary for policies to be durable enough to contribute to the disruption of the status quo. Unfortunately, the quest for naïve stability that results in accommodation with status quo interests can go wrong in just the way that the editors explain, and the carbon trap is at work here too.

16.3 The Double Trap

One temptation in the face of the inability of climate policies to survive long enough to catalyze decarbonization is to seek policies that won't be *too disruptive* or too politically contentious. Telling examples include the fractious debate over the proposed Green New Deal *within* the Democratic Party in the United States after President Joe Biden's election in 2020, and critiques of the European Green Deal for hollowing out its more transformative substantive as well as democratic, equity, and redistributive elements (Adler and Wargan 2022; Alemany 2021). Specific policies that focus on modestly reducing emissions while avoiding disruption to the carbon locked-in status quo are often the result. Such efforts aim to improve the carbon *efficiency* of systems or offer modest reforms without generating sufficient decarbonization or challenging dependence on fossil energy. Examples include energy efficiency programs, switching to natural gas as a lower-emission "bridge fuel," and carbon market schemes with prices much too low to generate real change in demand.

Such policies may be stable, but perniciously so, precisely because they are not political enough to challenge the status quo. This pattern suggests a double trap with the same kind of curve but with three equilibria: high carbon lock-in, more efficient carbon lock-in or improvement, and decarbonization (see Figure 16.2). Some climate policy efforts may generate emissions reductions but also generate dynamics that get systems stuck in improvement. Bridge fuel policies that promote natural gas as an alternative to coal will reduce emissions (assuming methane leakage is addressed). However, they may also entrench industrial interests that oppose moves toward deeper decarbonization, which cannot include natural gas (Betsill and Stevis 2016). Similarly, returning to our auto sector example, Toyota, once a leader in transitioning the sector with its innovation and production of hybrid technology, shifted to lobbying *against* EV mandates because it has become entrenched in hybrid production and has bet on hydrogen technology, which lags significantly compared to EV technology (Tabushi 2021). Depoliticized policy implementation strategies may yield similar patterns. For example, individual nudging efforts have been shown to decrease public support for broader policies such as carbon pricing, as people believe they have done enough (Hagman et al. 2019).

The very characteristics that allow such "improvement" policies to escape high carbon lock-in (depoliticized, technocratic, emissions-focused) contribute to their stability precisely because they do not challenge the status quo in a threatening manner. In a political context in which achieving significant climate action can be difficult, scholars have rightly focused on the necessary conditions for initiating action and building stability. However, the improvement trap indicates that we

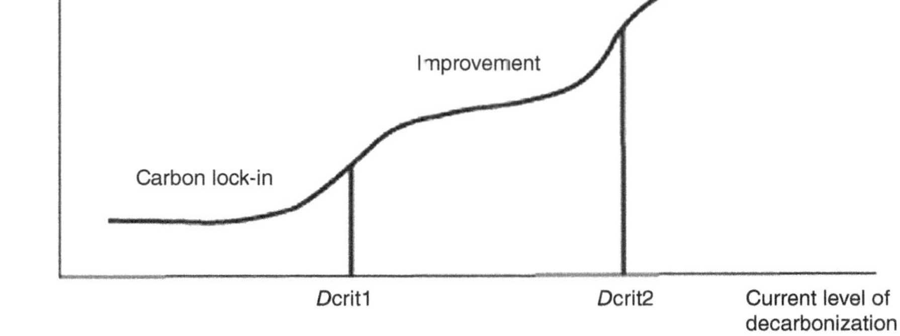

Figure 16.2 The double (improvement) trap.

should also give thought to how and under what conditions action can be ramped up following initial action and whether and how we can overcome the policy stability of the improvement trap.

16.4 Just Transitioning Out of the Carbon Trap

So far, our argument on how to reconcile the apparent politicization/disruption and stability tension has focused on putting politics at the center of analyses of why transformative policies have been so challenging to achieve, and how such policies might generate the politics to support stable trajectories. In this section, we highlight the pragmatic requirements and normative imperatives to produce the desired directionality. Here, repoliticization as understood by the editors – keeping or bringing climate action into political arenas where power, contestation, or competition is acknowledged and engaged – takes a more central role. We argue that just transition policies (Hughes and Hoffmann 2020; Indigenous Environmental Network 2017; Jasanoff 2018; McCauley and Heffron 2018; Newell and Mulvaney 2013; Newell et al. 2023) can add the right mix of politicization/disruption to promote stable trajectories in a more or less transformative direction, necessary to overcome the improvement trap and generate momentum toward decarbonization. We have three reasons for supporting this wager.

First, the empirical findings of our "politics of decarbonization" project (Bernstein and Hoffmann 2018, 2019) suggest that decarbonization initiatives that go beyond technological deployment to integrate justice, equity, and democracy concerns have more success scaling up and entrenching. This finding supports the case for entrenching goals that are explicitly justice-based, while allowing the necessary political processes to adapt and adjust. That will require significant political and normative work, and mobilization, up front. We already see several shifts in this regard that are a cause for optimism. Examples include the rise in youth mobilization driven by movements such as Fridays for Future (Fisher 2019), mobilization around the idea of just transitions even as its meaning remains contested (Stevis 2023; Wang and Lo 2021), proliferation of just transition policies and debates at local, national, and international levels (Hughes and Hoffmann 2020; Krawchenko and Gordon 2021; Newell et al. 2023), and more general normalization of the need for climate action.

While entrenching normative goals is essential through political processes, we should not naïvely believe that the ongoing political work needed will be without struggle or that entrenching openness and accountability in policy and political processes will be simple. The ongoing contestation over and criticism of the European Union's (EU) Just Transition Mechanism – arguably the world's most extensive and comprehensive attempt to implement the idea of just transition – is a case in point: Since its passing, social movements and academic critics have continued to press for change, arguing its social dimension is underdeveloped, it lacks stakeholder participation, and its funding is inadequate. And that is just a microcosm of ongoing contestation over the European Green Deal's broader contradictions and compromises, including its lack of consideration for distributional consequences or ecological harms (such as from mining for renewables) for countries and people outside the EU (Adler and Wargan 2022; Akgüç, Arabadjieva, and Galgóczi 2022).

Second, and following from this logic, just transition policies may provide a mix of politicized disruption and potential for stability. Efforts at improving justice and equity are inherently a political challenge to the status quo. Just transition thought and possible action are no exceptions. The concept is aimed at multiple areas of injustice related to climate change: injustice inherent to systems dominated by the extractive fossil energy sector (unequal distribution of the benefits and costs of fossil energy extraction); injustice in the distribution of the effects of climate impacts (those least responsible for causing the problem face the largest consequences); injustice in the distribution of costs/benefits of climate action itself (the very elements just transition policies are supposed to address, without which opposition is likely to grow on the left and right). Working in these areas challenges the status quo in multiple ways and underscores the inherent and multivalent injustice of the status quo.

Third, just transition is not just about disruption. It has the potential to (re)politicize the status quo in a critical way that offers a path toward new stable equilibria. Here, the reframing of the apparent paradox pays off and shifts to a pragmatic approach to transformation. If policy stability requires the generation of political/economic coalitions and underlying normative principles, just transition should have significant potential for generating stability. It identifies the "good" life in a low-carbon world. It also has the potential to activate and empower those marginalized by the carbon locked-in status quo and those potentially disrupted by decarbonization efforts, such as workers in the fossil fuel sector. More broadly, it has the potential to identify and connect winners from specific kinds of climate policies/actions in a way that could counter concentrated losers. In this way, just transition approaches could extend the kind of existential politics described by Colgan, Green, and Hale (2021) that occurs between owners of climate-vulnerable assets and climate-forcing assets. As our framework describes, the critical thresholds beyond which stability is desirable are made up not just of economic calculations but of the politics, norms, and institutions that shift perceptions of risks and benefits around these assets and expectations around the costs and benefits or societal acceptability of holding those assets in the future. Just transition policies have the potential to add a social dimension to this political conflict – for example by highlighting and possibly internalizing social costs of inaction and creating expectations around democracy, equity, and accountability – that could tip the balance of power toward stable, just low-carbon pathways.

16.5 Conclusion

The improvement trap we have identified highlights the importance of recognizing the risks of depoliticization and premature stability prior to reaching thresholds beyond which systems move toward the just decarbonization "attractor." Even then, when stability is desirable, our framework highlights the importance of attention to politics to sustain that trajectory. We recognize that identifying those thresholds in practice may be challenging while they are unfolding, but the carbon trap heuristic can highlight patterns to look for when designing or engaging with politics in practice.

One final example nearly perfectly illustrates the challenge: the pursuit of "net zero." The goal has proliferated, spurred by the 2015 Paris Agreement Article 4's articulation of the aim, albeit with several caveats, to "achieve a balance between anthropogenic emissions by sources and removals by sinks of greenhouse gases in the second half of this century." The goal is arguably now the "organizing principle" (Green and Reyes 2023) of climate politics. As of 2023, 150 countries, 254 cities, and almost half of the 2,000 largest companies in the world, which together account for 88 percent of global emissions and 89 percent of the world's population, have net zero policies or targets (Net Zero Tracker 2023). However, those impressive statistics mask the improvement trap.

"Net zero" pledges "solve" a wide range of political problems by ostensibly addressing climate change while creating a political consensus across left and right, thus being highly susceptible to depoliticization in pursuit of the goal. The trap becomes quickly apparent when examining the policies and consequences of net zero framing, which attracts a broad coalition precisely because it can support ongoing fossil fuel extraction and production, as we see prominently in cases like Canada's Pathways Alliance of oil sands producers.[3] This pattern is observable in a wide range of net zero policies, from massive investments in carbon capture technologies that effectively (or explicitly) subsidize ongoing and increasing extraction to policies that support decarbonization in some locations while exporting costs and social and ecological harms elsewhere, whether unsustainable land use changes, ecologically and socially harmful extraction, or economic pressures that will exacerbate inequalities and likely ensure continued carbon lock-in. Absent politicization, net zero can easily lead to "mitigation deterrence" resulting

[3] See the Pathways Alliance website: https://pathwaysalliance.ca/.

in significant overshoot of temperature targets and little attention to other negative sustainability impacts (Carton et al. 2023). Additionally, attempts to depoliticize can backfire, as opponents of climate action can just as easily politicize net zero as proponents, as we have seen with the mobilization of anti-climate populists against "net zero" (Paterson et al. 2023).

More generally, empirical and conceptual reviews of net zero policies and early implementation note a lack of attention to, and need for, politics (i.e. repoliticization) if net zero is to lead to just transformations toward climate goals (Carton et al. 2023; Fankhauser et al. 2021; Green and Reyes 2023). As we have argued, pragmatic and normative attention to just transitions are key elements of navigating these tensions and dilemmas.

Putting our own normative cards on the table, we are also simply not sure there is any other ethically defensible way forward than resolving the politicization/stability tension through a focus on just transitions, even as we recognize that not all good things always go together. Adler and Wargan (2022), for example, worry that the lesson so far of the EU's European Green Deal is that climate action can occur without democracy, or even undermine it. Others worry about climate authoritarianism (Mittiga 2022). Limping along with the false promise of stability in the improvement trap and low levels of climate action is also not a viable option. Fortunately, our own empirical work and the struggles we observe playing out in many places with the most active climate action suggest justice and necessary climate action are often aligned, and the politicization/stability debate suggests a normative imperative to ensure they continue to be (Patterson et al. 2018).

References

Adler, D. and Wargan, P. (2022). Decarbonization without democracy: Tennis-ball politics and the EU Green Deal. In K. Tienhaara and J. Robinson, eds., *Routledge Handbook on the Green New Deal*. London: Taylor & Francis, pp. 299–314.

Akgüç, M., Arabadjieva, K., and Galgóczi, B. (2022). Why the EU's patchy "just transition" framework is not up to meeting its climate ambitions. ETUI Policy Brief 2022.06. www.etui.org/publications/why-eus-patchy-just-transition-framework-not-meeting-its-climate-ambitions

Alemany, J. (2021). New York lawmaker rolls out Green New Deal for Public Schools. *Washington Post*, July 15. www.washingtonpost.com/politics/2021/07/15/climate-change-reconciliation-bill/

Allan, J., (2020). *The New Climate Activism: NGO Authority and Participation in Climate Change Governance*. Toronto: University of Toronto Press.

Auld, G., Bernstein, S., Cashore, B. et al. (2021). Managing pandemics as super wicked problems: Lessons from, and for, COVID-19 and the climate crisis. *Policy Sciences* 54(4), 707–728.

Bauer, M. W. and Knill, C. (2014). A conceptual framework for the comparative analysis of policy change: Measurement, explanation and strategies of policy dismantling. *Journal of Comparative Policy Analysis: Research and Practice* 16(1), 28–44.

Benzie, R. (2018). Tesla wins lawsuit against Ontario government over phase-out of electric vehicle incentives. *The Toronto Star*, August 18.

Bernstein, S. and Hoffmann, M. (2018). The politics of decarbonization and the catalytic impact of subnational experiments. *Policy Sciences* 51, 189–211.

Bernstein, S. and Hoffmann, M. (2019). Climate politics, metaphors and the fractal carbon trap. *Nature Climate Change* 9(12), 919–925.

Betsill, M. and Stevis, D. (2016). The politics and dynamics of energy transitions: Lessons from Colorado's "new energy economy." *Environment and Planning C* 24, 381–396.

Boisvert, N. (2019). Metrolinx removes electric vehicle chargers from GO station parking lots. *CBC*, January 10.

Breetz, H., Mildenberger, M., and Stokes, L. (2018). The political logics of clean energy transitions. *Business and Politics* 20, 492–522.

Bulkeley, H. A., Andonova, L. B., Betsill, M. M. et al. (2014). *Transnational Climate Change Governance*. Cambridge: Cambridge University Press.

Carton, W., Hougaard, I., Markusson, N., and Lund, J. (2023). Is carbon removal delaying emission reductions? *WIREs Climate Change* 14(4).

Colgan, J. D., Green, J., and Hale, T. (2021). Asset revaluation and the existential politics of climate change. *International Organization* 75(2), 586–610.

Fankhauser, S., Smith, S. M., Allen, M. et al. (2021). The meaning of net zero and how to get it right. *Nature Climate Change* 12(1), 15–21.

Fisher, D. R. (2019). The broader importance of #FridaysForFuture. *Nature Climate Change* 9, 430–431.

Green, J. and Reyes, R. S. (2023). The history of net zero: Can we move from concepts to practice? *Climate Policy* 23(7), 901–915.

Hagmann, D., Ho, E. H., and Loewenstein, G. (2019). Nudging out support for a carbon tax. *Nature Climate Change* 9, 484–489.

Hochstetler, K. (2020). *Political Economies of Energy Transition: Wind and Solar Power in Brazil and South Africa*. Cambridge: Cambridge University Press.

Hoffmann, M. (2011). *Climate Governance at the Crossroads: Experimenting with a Global Response after Kyoto*. New York: Oxford University Press.

Hughes, S. and Hoffmann, M. (2020). Just urban transitions: Toward a research agenda. *Wiley Interdisciplinary Reviews: Climate Change* 11(3), e640.

Indigenous Environmental Network. (2017). Indigenous Principles of Just Transition. www.ienearth.org/justtransition/

Jasanoff, S. (2018). Just transitions: A humble approach to global energy futures. *Energy Research and Social Science* 35, 11–14.

Jordan, A. and Matt, E. (2014). Designing policies that intentionally stick: Policy feedback in a changing climate. *Policy Sciences* 47(3), 227–247.

Jordan, A. and Moore, B. (2020). *Durable by Design? Policy Feedback in a Changing Climate*. Cambridge: Cambridge University Press.

Krawchenko, T. A. and Gordon, M. (2021). How do we manage a just transition? A comparative review of national and regional just transition initiatives. *Sustainability* 13, 6070.

Kuzemko, C. (2016). Energy depoliticization in the UK: Destroying political capacity. *British Journal of Politics and International Relations* 18(1), 107–124.

Levin, K., Cashore, B., Bernstein, S., and Auld, G. (2012). Overcoming the tragedy of super wicked problems: Constraining our future selves to ameliorate global climate change. *Policy Sciences* 45(2), 123–152.

Levy, D. L. and Newell, P. J., eds. (2005). *The Business of Global Environmental Governance*. Cambridge, MA: MIT Press.

Lockwood, M., (2022). Policy feedback and institutional context in energy transitions. *Policy Sciences* 55(3), 487–507.

McCauley, D. and Heffron, R. (2018). Just transition: Integrating climate, energy and environmental justice. *Energy Policy* 119, 1–7.
Meckling, J. (2019). Governing renewables: Policy feedback in a global energy transition. *Environment and Planning C*. 37, 317–338.
Millar, H., Bourgeois, E., Bernstein, S. and Hoffmann, M. (2021). Self-reinforcing and self-undermining feedbacks in subnational climate policy implementation. *Environmental Politics* 30(5), 791–810.
Mittiga, R. (2022). Political legitimacy, authoritarianism, and climate change. *American Political Science Review* 116(3), 998–1011.
Moore, B. and Jordan, A. (2020). Disaggregating the dependent variable in policy feedback research: An analysis of the EU emissions trading system. *Policy Sciences* 53(2), 291–307.
Mullaney, T. (2023). A key test for EV sales and the adoption curve is coming. *CNBC*, July 30. www.cnbc.com/2023/07/30/a-key-test-for-ev-sales-and-the-adoption-curve-is-coming.html.
Net Zero Tracker. (2023). Net Zero Tracker. https://zerotracker.net/.
Newell, P., Daley, R., Mikheeva, O., and Peša, I. (2023). Mind the gap: The global governance of just transitions. *Global Policy* 14(3), 425–437.
Newell, P. and Mulvaney, D. (2013). The political economy of the "just transition." *The Geographical Journal* 179(2), 132–140.
Paterson, M., Tobin, P., and VanDeveer, S.D. (2022). Climate governance antagonisms: Policy stability and repoliticization. *Global Environmental Politics* 22(2), 1–11.
Paterson, M., Wilshire, S., and Tobin, P. (2023). The rise of anti-net zero populism in the UK: Comparing rhetorical strategies for climate policy dismantling. *Journal of Comparative Policy Analysis: Research and Practice* 26(3–4), 332–350. https://doi.org/10.1080/13876988.2023.2242799
Patterson, J. (2023). Backlash to climate policy. *Global Environmental Politics* 23(1), 1–23.
Patterson, J., Thaler, T., Hoffmann, M. et al. (2018). Political feasibility of 1.5C societal transformations: The role of social justice. *Current Opinion in Environmental Sustainability* 31, 1–9.
Rabe, B. G. (2018). *Can We Price Carbon?* Cambridge, MA: MIT Press.
Rabson, M. (2023). Canadian electric vehicle registrations inching closer to liberal government's target. *Globe and Mail*, August 3. www.theglobeandmail.com/canada/article-canadian-electric-vehicle-registrations-inching-closer-to-liberal/
Rosenbloom, D., Meadowcroft, J., and Cashore, B., (2019). Stability and climate policy? Harnessing insights on path dependence, policy feedback, and transition pathways. *Energy Research and Social Science* 50, 168–178.
Stevis, D. (2023). *Just Transitions: Promise and Contestation*. Cambridge: Cambridge University Press.
Stokes, L. C. (2013). The politics of renewable energy policies: The case of feed-in tariffs in Ontario, Canada. *Energy Policy* 56, 490–500.
Stokes, L. C. (2020). *Short Circuiting Policy: Interest Groups and the Battle over Clean Energy and Climate Policy in the American States*. New York: Oxford University Press.
Tabushi, H. (2021). Toyota led on clean cars. Now critics say it works to delay them. *New York Times*, July 26 (updated August 6). www.nytimes.com/2021/07/25/climate/toyota-electric-hydrogen.html
Unruh, G. C. (2000). Understanding carbon lock-in. *Energy Policy* 28(12), 817–830.
Wang, X. and Lo, K. (2021). Just transition: A conceptual review. *Energy Research and Social Science* 82, 102291.

17

Conclusions for Stability and Re/politicization in Climate Governance

STACY D. VANDEVEER, PAUL TOBIN, AND MATTHEW PATERSON

Contemporary political life is bewilderingly contradictory. In 2022, the six biggest oil companies doubled their annual profits to just over USD 200 billion. Meanwhile, December 2023 saw the conclusion of the 28th Conference of the Parties (COP) to the United Nations Framework Convention on Climate Change (UNFCCC). For the first time at the COPs, there was recognition of the need for "transitioning away from fossil fuels in energy systems." This statement was not as ambitious as many wanted: "recognition of" is not as ambitious as a "commitment to," while "transitioning away" does not equate to "phasing out." Indeed, the winding path to securing this latest and underwhelming "landmark" document in December 2023 epitomizes the rationale behind this book. To date, humanity has struggled with how to navigate toward a transformation in our relationship with greenhouse gas (GHG) production over a multi-decade time period. Meanwhile, numerous assessments concluded that 2023 was, by far, the hottest year on record, with many significant climate "anomalies" (NOAA 2023; Schmidt 2024). The year 2023 also saw record-high global GHG emissions (IEA 2024). Frankly, contemporary political life is extremely worrying, as well as bewilderingly contradictory.

The motivations for this book arose from a simple observation with complex and dynamic implications: As a political and policymaking issue, climate change requires long-term action and significant institutional change to transform our current environmental, economic, and governance systems yet faces a wide array of powerful and influential vested interests that are opposed to many of the changes required. Exactly how we achieve significant and transformational change remains deeply contested.

Within academic scholarship on securing long-term action, we see a clear antagonism between two persuasive yet seemingly mutually exclusive schools of thought (Paterson et al. 2022). On the one hand, the pursuit of *stability* in policymaking and governance appears intuitive for addressing a decades-long challenge: maintain, or even lock in, patterns of action that may, for example, constrain future behavior or self-reinforce to create coalitions of facilitators that continue activity (see Roberts et al. 2018; Rosenbloom et al. 2019). Yet scholars and practitioners have warned that such approaches bring their own difficulties, given the considerable influence of agents and institutions that wield the most power. At its most stark, opponents to climate action have funded lobbyists to oppose basic science, obstruct new proposals, and dismantle those that make it into law (Dunlap and Brulle 2020; Stokes 2020). Critics warn that an emphasis on stability that does not, or cannot, counteract these influences simply perpetuates them. In response, calls for *re/politicization* have placed conflict and disruption at the heart of their strategies (Allan 2020; Kenis and Mathijs 2014), such that the vested interests opposing climate action over recent decades, particularly the highly resourced fossil fuel companies, are explicitly opposed. From this perspective, any approach that pursues stability while such influential actors hold the means to block ambition to protect their profits will, by definition, be inadequate.

In response to this antagonism between stability and re/politicization, we brought together twenty-six scholars to explore eleven country cases, and multiple policy areas, across governance contexts ranging from the local to global scale. We began, in Part I, by examining *movement politics*, as this is where much of the discussion around the need for re/politicization has arisen. In Part II, we turned our attention to the importance of *political economy*, as the industries we analyze have often been the subject of opposition for social movements, as well as commonly being well-funded centers of resistance and opposition to the political and economic changes needed to meet the challenges posed by climate change. In Part III, contributors' chapters focused on *comparative political dynamics*, to see how these tensions play out in different countries across the world. Part IV shifted our attention to how the tensions play out in *global politics*, specifically with the UNFCCC and global trade governance. Finally, in Part V, our *reflections*, we critiqued the binary portrayal of stability and re/politicization.

We begin this concluding chapter by outlining what we have learned regarding stability from our chapters. We then do likewise for re/politicization. Third, we reflect on the importance of temporality for our analyses. Next, we turn to depoliticizing strategies and discourses. We then emphasize the importance of justice before finally offering concluding remarks and suggesting future avenues for research.

17.1 Understanding Stability

In Chapter 1, we identified four forms of stability: stability as the *status quo*, stability as engineering lock-in, stability as policy lock-in, and stability as long-term emissions reduction pathways. What have the chapters in between taught us about these forms of stability in climate politics?

First, there are many situations where the pursuit of stability in terms of engineering lock-in to generate long-term emissions reductions, or simply as stable policy environments, in practice generates stability as the *status quo*. While focusing on the Fridays for Future movement in Germany, Tosun and Debus (Chapter 2) argue that the steady growth of climate policies may face dismantling strategies, unless some degree of stability is pursued. Méndez (Chapter 3) shows how climate policy framed in terms of what he calls "carbon reductionism" narrows the scope of governance to one that is manageable within existing power arrangements at the expense of climate justice. And Tobin, Ali, MacGregor, and Ahmad (Chapter 4) show that stable policy environments can entrench existing racialized power hierarchies but also that the existence already of such hierarchies necessarily makes marginalized communities wish to avoid conflict and contestation for fear of greater repercussions. Using the case of peatland restoration, Paterson (Chapter 6) shows that as climate policy expands into new domains, the existing power relations within those domains limit the ability of climate policy to reshape them effectively. Chapter 7 by Stephens and Sokol and Chapter 8 by Haufler both show that in relation to the financial sector, actors tend to get "captured," resulting in the pursuit of financial stability as opposed to aggressive emissions reductions.

The chapters show that a key question that arises is "stability for whom?" A stable policy environment inevitably works to the advantage of some and the disadvantage of others. Stephens and Sokol make this point most forcefully in the context of central bank climate activity, but it applies similarly to questions of climate justice (Méndez) and the participation of Muslim communities, as well as, by extension, other marginalized groups (Tobin, Ali, MacGregor, and Ahmad).

But other chapters complicate this picture somewhat. VanDeveer (Chapter 5) suggests that politicizing actors seek a form of stable, policy lock-in, as they work to phase out and permanently ban coal burning via universalizing anti-coal burning and anti-coal financing policy norms. Even when climate policy is depoliticized, and focuses on stable climate policy over time, there are important contexts where it is able, in practice, to generate fairly ambitious climate action. For example, Hochstetler's contribution (Chapter 9) demonstrates this situation powerfully in her comparison of Brazil and South Africa, where depoliticized contexts and policy frameworks in Brazil are able to ramp up wind energy considerably, whereas highly politicized energy politics in South Africa works to impede energy transitions there. Sun, Shen, and Lewis (Chapter 10) find that the Chinese state is able to develop climate policy in ways that appear oriented toward stable policy environments, but at various points in time, this entails strategically organizing conflicts with incumbent interests to undermine their ability to block climate action. Farstad, Hermansen, and Lahn (Chapter 11) identify several important successes of depoliticized climate policy interventions in Norway, underscoring their emphasis on the need for agnosticism regarding the antagonism between stability and conflict. Torney (Chapter 12) shows that even while policymakers seek to generate stable policy over time, the framework climate laws they generate do so in ways that seek to undermine the status quo, destabilizing existing power relations to generate more aggressive climate action. Similarly, Mildenberger and Lockwood (Chapter 13) argue that the "depoliticized" character of prominent climate institutions is often misunderstood, as they arise out of political contestation and have been designed in recognition of the political conflicts inherent to climate action.

Our latter chapters provide a critical lens for reflecting on stability. Allan (Chapter 14) builds on her previous work outlining the "dangerous incrementalism" of the UNFCCC system (see Allan 2019) through her analysis of climate action at the global level, in which she finds stability to be fragile and difficult to maintain. Nahm's chapter on US-China relations and green industrial policy (Chapter 15) explores how politicization could be used *in order to achieve* stability but that in doing so it risks undermining the existing global economic order and the foundations for long-term climate collaboration. Building neatly on this interlinked relationship between politicization and stability, Bernstein and Hoffmann (Chapter 16) argue – and consolidate the points made by Farstad, Hermansen, and Lahn and Mildenberger and Lockwood, via their analysis of the "carbon trap" – that these two concepts are not dichotomous, and instead that pursuing stability is an inherently political process.

17.2 Understanding Politicization and Repoliticization

What do our contributors have to say about the four forms of re/politicization we identified in Chapter 1? We name these forms: politicization as pursuit of broader sociopolitical change; politicization as pursuit of partisan competition; politicization as rhetoric; and politicization as scholarly praxis. First, and perhaps most obviously, considering the origins of politicization activities, we see multiple examples of politicization emerging from explicit contestations by social movements. Indeed, the opening three-chapter Part I of our book, on *movement politics*, analyzes that topic directly. Our chapters in Part I assess this pursuit of politicization via the impacts of the Fridays for Future movement on the coal phaseout in Germany (Tosun and Debus), the ability of climate justice movements in California to contest "carbon reductionism" and broaden the scope of climate action in the state (Méndez), and Muslim communities' challenging dilemma as to whether the pursuit of politicization creates risks for campaigners. Subsequent chapters build on these themes – for example, the chapter by Farstad, Hermansen, and Lahn, which discusses how various NGOs have effectively targeted particular tensions within Norwegian policy, notably over gas-fired power plants and carbon capture and storage (CCS), and the efforts of NGO campaigners to generate more ambitious climate action by insurers, in Haufler's chapter. VanDeveer's chapter draws attention to domestic and transnational activists – along with self-declared lead states – that are seeking to expand the list of states, subnational public sector units, and financial institutions declaring fossil fuel phaseout deadlines, or opposing the financing of new coal infrastructure. In some contexts, these activities entail the pursuit of broader sociopolitical change, while in others it is to effect more narrow gains in climate policy action. Sometimes, the logic of movement activism is more complex, with competing focuses of campaign activity across environmentalists, as in Paterson's chapter on the differing uses and values of peatlands around which campaigns are organized.

But some chapters show that politicization comes from less obvious sources. In both Allan's and Nahm's chapters, highly powerful actors – states, especially dominant ones in global politics – repoliticize existing international agreements to pursue their interests (i.e. the United States regarding the Kyoto Protocol), or politicize the domestic climate policy of other states within other international fora, such as the World Trade Organization. Sun, Shen, and Lewis similarly show that politicization is sometimes pursued by the Chinese state (even within a broadly depoliticized policy regime) in order to secure compliance from particular companies or other actors. Conversely, other chapters show that politicization can generate dangers for marginalized actors, such as Muslim climate actors in the UK (Tobin, Ali, MacGregor, and Ahmad), for reasons such as unequal treatment via the justice system. But politicization can also generate dangers for undermining climate policy itself: As Nahm argues, seeking to onshore manufacturing in the United States for renewable energy technologies might build coalitions for climate policy in the United States, but it also raises costs and could slow down the rate of US solar installation. VanDeveer's contribution argues that the movement to universalize coal phaseout policies has met, and sometimes engendered, politicized distributional and procedural justice debates at the local and global scale.

Many of our chapters entail politicization as an academic activity as well as an empirical focus. This scholarly praxis is usually pursued through normative arguments that propose the politicization of various aspects of climate policy – for example, politicizing central banks (Stephens and Sokol), the financial sector in general (Haufler), fossil fuel incumbents (Hochstetler; VanDeveer), or through rhetorical strategies such as just transitions (Bernstein and Hoffmann). Other chapters are less immediately normative but nevertheless assume that even if we do not *yet* see much politicization of a given aspect of climate policy, it is immanent to the logic of that policy domain (see, e.g., Paterson on peatlands). Indeed, as discourses and activism associated with justice, injustice, and anti-fossil fuel norms (Blondeel et al. 2019; Mitchell and Carpenter 2019) are globalized, it is difficult to imagine what a related un-politicized or normatively stable climate change politics might look like.

17.3 Combining Stability and Politicization

A running thread through the book is that stability and politicization are not dichotomous strategies for policymakers or others to pursue, even if they appear to be competing approaches to climate action. In practice, they interact considerably in ways that are irreducible to the overall logic of "accomplishing" climate governance (Bulkeley 2015). Both Torney's analysis of framework climate laws and Mildenberger and Lockwood's of climate policy institution-building show that, contrary to the assumption that these processes are intended to be depoliticizing, they have the possibility of repoliticization built into them. Bernstein and Hoffmann complement this argument about how we understand climate policy and institutional design with a model of decarbonization over time, showing that the logics of capture we have seen already (how stability in policy design can become stability as the *status quo*) can generate semi-transitions to situations they call "improvement," but this situation then is likely to generate additional politicizations that renew pressure for additional action. VanDeveer's chapter demonstrates that politicizing coal burning, in pursuit of more significant and faster climate change mitigation via permanently banning coal burning and coal sector financing, can itself be repoliticized via justice and decolonial concerns at the local and global scale.

But for Mildenberger and Lockwood, and for Torney, it is not only academics who argue that the logic of climate policy means that stability-oriented policy regimes cannot be sustained over time. Certainly decolonial scholarship is often rather explicit about the need to politicize scholarship as well as broader political and social institutions. Beyond this "politicization as scholarly practice," policymakers themselves have also designed institutions that enable future political conflict over climate change, recognizing the need for both elements in our framework to be integral to future climate strategy.

17.4 Depoliticizing Strategies and Discourses

Many of our chapters show that attempts to pursue stable policy regimes are very difficult to sustain over time, and that such attempts periodically unravel. They also show that in practice many such policy regimes are less *depoliticized* (Feindt et al. 2021; Wood 2016) than often assumed, containing political bargains within them and at times designed to reshape power relations in favor of further decarbonization. Nevertheless, the urge for policymakers and many researchers to engage in depoliticizing strategies and discourses is powerful. Pursuit of depoliticization often arises out of the identification by policymakers with particular incumbent interests, which makes them seek to elide or obscure the contradictions in climate policy that result from these identifications. But it can also arise more simply because the normative rhetoric of "we are all in this crisis together" can be an effective discursive strategy for bringing diverse actors together around climate policy, including actors who may not otherwise choose to participate if engagement is – or appears to be – more "political" than they would wish. In our introductory chapter, we unpacked the term "politics" into its different elements. We deployed three such elements – politics as a site of collective decision-making, politics as power, and politics as conflict (see Paterson 2021). Resultantly, we can understand these depoliticizing urges precisely as the effects of power and conflict on how collective decision-making is made. Depoliticization strategies are therefore deeply political in their origins and effects.

We see the urge to present climate policy as depoliticized in various ways in our chapters. Tosun and Debus show that the politicization of climate change by Fridays for Future, and with the Green Party as a government coalition partner, threatened to undermine long-term climate policy and generated incentives for governing actors to depoliticize it in response. This effect has in part been generated by the nature of the German institutional framework, an effect that is even more obvious in Norway (see Dryzek et al. 2002), where Farstad, Hermansen, and Lahn show that deeply institutionalized norms about consensual decision-making produce these incentives for depoliticizing strategies. And while analyzing four countries (the UK, Australia, Norway, and Denmark), Mildenberger and Lockwood unpack neatly the political bargains involved in generating "stability-oriented" climate institutions. Yet they nevertheless show that such institutions are routinely presented rhetorically by policymakers and politicians as being "outside" partisan politics.

Stephens and Sokol, as well as Haufler, show the core rationale for depoliticization even more starkly. The powerful financial sector not only organizes and frames climate governance in depoliticized ways to avoid scrutiny of its activities in relation to climate change but also uses climate policy to advance its interests directly. Similar dynamics are detailed in Hochstetler's chapter, whereby the complex and contentious politics around forests in Brazil incentivizes governments to avoid conflicts and focus on less conflictual areas of climate action, notably wind energy.

But while the incentives for *dominant* actors to engage in depoliticizing strategies are increasingly acknowledged and understood, Tobin, Ali, MacGregor, and Ahmad show that this can also be the case for marginalized actors. In their chapter, they show that many Muslim climate actors in the UK feel unable to participate in more contentious climate action because of the dynamics of structural or institutional racism, and hence often pursue more depoliticized strategies for reasons of personal security. In other contexts, depoliticization strategies arise out of the habitus of the more technically focused actors in many parts of climate governance. For example, the scientists who are involved in peatland management (Paterson) often focused on how to restore peatlands to minimize carbon losses (and even absorb carbon), rather than political questions around land ownership. Indeed, spending time on such topics would feel like a dangerous distraction.

Lastly, several chapters demonstrate that what depoliticizing actors are trying to avoid – and/or what they perceive as problematic about "politics" – also various substantially. For Stephens and Sokol, regarding the treatment of central banks, and for Haufler's contribution on the insurance and reinsurance sector, we see actors who generally seek to keep contentious noneconomic politics "out" of their economic governance discourses. Paterson's chapter on "peaty politics" similarly argues that many actors want to keep contentious areas of climate change politics out of peat governance. In all three of these contributions, highly technical discourses and bodies of knowledge are deployed to insulate or protect governance decision-making process from contentious or more explicitly normative aspects of politics. Certainly, there is a long tradition of such uses of science and technology in many areas of environmental politics (see, e.g., Jasanoff 1994). Meanwhile, for Farstad, Hermansen, and Lahn, depoliticizing Norwegian climate policy was, for many years, accomplished through the invocation and maintenance of societal and cross-political-party consensus that shows more recent signs of fraying.

17.5 Understanding Antagonisms and Iterated Politics over Time

If the urge to *de*politicize climate policy in the pursuit of stable climate policy regimes is strong for many actors, so too is the desire in others to *re*politicize. In part, this desire is integral to the logic of climate change itself: For many, there is a deep sense that attempts to depoliticize climate change – in the sense of presenting it as outside or beyond power relations and conflicts in society – is simply mistaken as both a strategy and an understanding of societal dynamics. From this perspective, depoliticizing strategies in climate policy are always in some sense merely tactical rather than coherent intellectually. However, this point means that the urge to repoliticize arises out of direct awareness of those broader power relations and conflicts in society. Particular groups of people have reasons to mobilize over climate change. Our chapters illustrate these actors in various ways – from youth-centered movements (Fridays for Future – see Tosun and Debus) or other climate justice groups (Bernstein and Hoffmann) to those affected negatively by either climate policy or climate change itself (see chapters variously by Méndez; Farstad, Hermansen, and Lahn; or Tobin, Ali, MacGregor, and Ahmad). Academics are of course actors in climate politics too, and several of our chapters (Bernstein and Hoffmann's argument for just transitions, Méndez's arguments about "climate change from the streets," and Stephens and Sokol's call to politicize central banking) can be read as repoliticizing interventions, as indeed can the book as a whole.

We need to see the relationships between stability and repoliticization as iterated over time. Each contains logics that arise out of the urges of some sets of political actors to frame and act on climate in specific ways, which then provoke reactions from other actors to frame and act on these differently. The institutionalization of stability-oriented climate policy also contains contradictory logics that tend to falter and create repoliticizing moments. Conversely, opening up new lines of political conflict risks opening up climate action to interventions by those pursuing a backlash to, or dismantling of, climate policy (Patterson 2023; Paterson et al. 2024), generating new rounds of attempts to "lock in" climate policy to prevent such backsliding.

How you – a particular reader of this book – respond to these dynamics depends on your positionality within the various processes we analyze here. For many, activism that repoliticizes climate seems like the obvious option due to lack of influence within the current structures. This obviousness results from one's own standpoint. For example, from a generational perspective, as many readers of the book will be students (and thus commonly young adults), activism is one of the few available opportunities for action and feeds off intergenerational resentments and conflicts that are intrinsic to the temporal dynamics of climate change. Climate change is indeed a phenomenon created by older, richer (white) people in the Global North, and imposed on the young everywhere, albeit highly unequally young people across the world. There is also a sense among many young students and activists that we are all running out of time to avoid catastrophic impacts (de Moor 2023), and that these will occur over the expected lifespans of today's younger generations.

But for other readers, those with perhaps more of a sense of access – if mediated – to political decision-making, and perhaps also more to lose having established careers and families, or indeed, those with concern for unfair treatment in justice systems, such activism may seem inaccessible or unappealing. For these readers, the pursuit of climate action via political parties, lobbying, organized interest groups, or professional networks, which favor less conflictual action, may be more in line with their situation. This approach is not necessarily depoliticized – it would include partisan competition among political parties – but it is nevertheless not "climate change from the streets," in Méndez's words. Hence, it does contain the seeds of a more stability-oriented approach to climate change, seeking consensus where possible.

17.6 Concluding Remarks and Future Research

Our aim in this book was to provide an analytical framework based around the antagonisms between "stability" and "politicization" as a means for thinking about the political dynamics of long-term climate policy. We think the framework is useful empirically in that it captures common empirical phenomena within climate policy – both the urge to seek to "lock in" climate policy to pursue decarbonization in a way that prevents backsliding and the urge to open up political conflict that can destabilize existing power relations blocking ambitious climate policy action. And our framework is also useful normatively – both poles in our framework contain powerful arguments that motivate actors to act on climate change in specific ways, and hence clarifying what is at stake in these two approaches proffers analytical benefits. Moreover, it is important to recognize the normative value of both logics, and especially the value of the empirical relationships between them.

We have presented the framework in a deliberately open and flexible manner, and we hope others will find it useful in developing their own analyses. We particularly hope the four forms of stability and the four forms of politicization that we identify through this volume can serve to make future research in these fields even more nuanced. Indeed, these forms can be employed in myriad ways. Some may simply see our antagonist framework, and its attendant eight forms, as useful means to organize their own case study or comparative research. Others may prefer to develop the approach into a more formal model that might be constructed and tested.

While we have sought to maximize the diversity of empirical contexts in which the arguments have been developed, there are nevertheless many ways they could be evaluated and developed in other empirical contexts. For example, how does this antagonism play out in cities? The chapters by Allan and by Nahm extend the argument beyond the national policy level that dominates existing literature, but there is much more to be said about stability and repoliticization at inter-state and transnational spaces, where relationship dynamics are necessarily orchestrated differently than in national, subnational, and non-state arenas. Cities hold the potential to play a unique role in climate governance yet possess unique dynamics compared to other levels (van der Heijden 2019) and have not been analyzed in this volume. Similarly, what about the important sectors that we lacked the space to explore here? For instance, we have only one chapter here dealing with transport as part of its analysis, and which pertains to electric vehicle (EV) transitions, while there is nothing on agriculture and food or aviation and very little on buildings. Each of these sectors likely possess their own actor coalitions, priorities, conflicts, and focal issues (see Schmidt et al. 2022). It would be useful to test the arguments proposed in this book in various contexts to see how they hold up or need adapting according to their context. Furthermore, as we note in our introductory chapter, the bulk of our analyses pertain to climate mitigation rather than adaptation, yet the pursuit of long-term and effective policy, and the dynamics around how to do so, pertain as much to responding to the impacts of climate change as to its prevention. Further research into the antagonisms that underpin adaptation policy, and the inequalities of power that influence decision-making in this regard, is of the utmost importance and urgency to those regions already facing the harshest effects of climate change, often despite playing a limited role in their creation.

There is a lot at stake in getting our understanding of these dynamics right. We are finalizing this book in 2024, which is a year of a huge number of elections around the world: Amy Davidson Sorkin, writing in the *New Yorker*, dubbed it "the biggest election year in history" (Sorkin 2024), with perhaps seventy-six countries with universal suffrage going to the polls, comprising more than 40 percent of the world's population. The United States, the UK, Mexico, South Africa, India, Russia, Pakistan, and Iran are just some of the most prominent national elections for presidencies and/or parliaments to come during the year – a year that also includes European Parliamentary elections. Climate change is once again a partisan issue in many contexts, including those, like the UK, where it has until recently enjoyed bipartisan support (Carter and Pearson 2024). Many of these elections could destabilize the political conditions for climate policy in dramatic ways. And beyond electoral politics, there remains ongoing mobilization both to accelerate climate policy action and to undermine – or even dismantle – it. These activities are likely to put considerable pressure on existing climate initiatives around the world, and in particular on the institutionalized forms through which a previous round of climate policymaking has sought to "lock in" climate action over time. How these challenges play out empirically, and how different actors are able to use them not only to withstand the pressure for climate policymaking but perhaps also to intensify and accelerate climate action, is crucial for the future of the global response to the climate crisis.

References

Allan, J. I. (2019). Dangerous incrementalism of the Paris Agreement. *Global Environmental Politics* 19(1), 4–11.

Allan, J. I. (2020). *The New Climate Activism: NGO Authority and Participation in Climate Change Governance*. Toronto: University of Toronto Press.

Blondeel, M., Colgan, J. and Van de Graaf, T. (2019). What drives norm success? Evidence from anti-fossil fuel campaigns. *Global Environmental Politics* 19(4), 63–83.

Bulkeley, H. (2015). *Accomplishing Climate Governance*. Cambridge: Cambridge University Press.

Carter, N. and Pearson, M. (2024). From green crap to net zero: Conservative climate policy 2015–2022. *British Politics* 19(1), 154–174.

de Moor, J. (2023). Time and place in climate activism: Three urgency-induced debates. In J. Sowers, S. D. VanDeveer and E. Weinthal, eds., *Oxford Handbook of Comparative Environmental Politics*, Oxford: Oxford University Press, pp. 299–316.

Dryzek, J. S., Hunold, C., Schlosberg, D., Downes, D. and Hernes, H. K. (2002). Environmental transformation of the state: The USA, Norway, Germany and the UK. *Political Studies* 50(4), 659–682.

Dunlap, R. E. and Brulle, R. J. (2020). Sources and amplifiers of climate change denial. In D. C. Holmes and L. M. Richardson, eds., *Research Handbook on Communicating Climate Change*, Cheltenham: Edward Elgar, pp. 49–61.

Feindt, P. H., Schwindenhammer, S. and Tosun, J. (2021). Politicization, depoliticization and policy change: A comparative theoretical perspective on agri-food policy. *Journal of Comparative Policy Analysis: Research and Practice* 23(5–6), 509–525.

IEA (International Energy Agency). (2024). *CO_2 Emissions in 2023: A New Record High, but Is There Light at the End of the Tunnel?* Report. www.iea.org/reports/co2-emissions-in-2023

Jasanoff, S. (1994). *The Fifth Branch: Science Advisors as Policymakers*. Cambridge, MA: Harvard University Press.

Kenis, A. and Mathijs, E. (2014). Climate change and post-politics: Repoliticizing the present by imagining the future? *Geoforum* 52, 148–156.

Mitchell, R. B. and Carpenter, C. (2019). Norms for the Earth: Changing the climate on "climate change." *Journal of Global Security Studies* 4(4), 413–429.

NOAA (National Oceanic and Atmospheric Agency). (2023). *Annual 2023 Global Climate Report*. www.ncei.noaa.gov/access/metadata/landing-page/bin/iso?id=gov.noaa.ncdc:C00672

Paterson, M. (2021). *In Search of Climate Politics*. Cambridge: Cambridge University Press.

Paterson, M., Tobin, P. and VanDeveer, S. D. (2022). Climate governance antagonisms: Policy stability and repoliticization. *Global Environmental Politics* 22(2), 1–11.

Paterson, M., Wilshire, S. and Tobin, P. (2024). The rise of anti-net zero populism in the UK: Comparing rhetorical strategies for climate policy dismantling. *Journal of Comparative Policy Analysis: Research and Practice* 26(3–4), 332–350.

Patterson, J. J. (2023). Backlash to climate policy. *Global Environmental Politics* 23(1), 68–90.

Roberts, C., Geels, F.W., Lockwood, M. et al. (2018). The politics of accelerating low-carbon transitions: Towards a new research agenda. *Energy Research & Social Science* 44, 304–311.

Rosenbloom, D., Meadowcroft, J. and Cashore, B. (2019). Stability and climate policy? Harnessing insights on path dependence, policy feedback, and transition pathways. *Energy Research & Social Science* 50, 168–178.

Schmidt, G. (2024). Climate models can't explain 2023's huge heat anomaly. *Nature*, 627, 427.

Schmidt, N. M., Tobin, P. and Moore, B. (2022). Brace for turbulence: EU member states' climate strategies in the aviation sector. *Climate Action* 1(1), 1–7.

Sorkin, A. D. (2024). The biggest election year in history. *The New Yorker*, January 7. www.newyorker.com/magazine/2024/01/15/the-biggest-election-year-in-history.

Stokes, L. C. (2020). *Short Circuiting Policy: Interest Groups and the Battle over Clean Energy and Climate Policy in the American States*. New York: Oxford University Press.

van der Heijden, J. (2019). Studying urban climate governance: Where to begin, what to look for, and how to make a meaningful contribution to scholarship and practice. *Earth System Governance* 1, 100005.

Wood, M. (2016). Politicization, depoliticization and anti-politics: Towards a multilevel research agenda. *Political Studies Review* 14(4), 521–533.

Index

Printed by Integrated Books International,
United States of America